*Techniques in free
radical research*

LABORATORY TECHNIQUES IN BIOCHEMISTRY AND MOLECULAR BIOLOGY

Volume 22

Edited by

R.H. BURDON—*Department of Bioscience and Biotechnology, University of Strathclyde, Glasgow*
P.H. van KNIPPENBERG — *Department of Biochemistry, University of Leiden, Leiden*

ELSEVIER
AMSTERDAM · LONDON · NEW YORK · TOKYO

TECHNIQUES IN FREE RADICAL RESEARCH

Catherine A. Rice-Evans

Anthony T. Diplock

Division of Biochemistry,
United Medical & Dental Schools of
Guy's & St. Thomas's Hospitals,
University of London, London, U.K.

Martyn C.R. Symons

Department of Chemistry,
University of Leicester, Leicester, U.K.

1991
ELSEVIER
AMSTERDAM · LONDON · NEW YORK · TOKYO

ISBN 0-444-81314-4 (pocket edition)
ISBN 0-444-81304-7 (library edition)
ISSN 0-7204-4200-1 (series)

Published by:
ELSEVIER SCIENCE PUBLISHERS BV
P.O. BOX 211
1000 AE AMSTERDAM
THE NETHERLANDS

Sole distributors for the USA and Canada:
ELSEVIER SCIENCE PUBLISHING COMPANY, INC.
655 AVENUE OF THE AMERICAS
NEW YORK, NY 10010
USA

Library of Congress Card No. 85-647011

Printed in the Netherlands on acid free paper

Preface

Free radical species are generally short-lived due to their high reactivity and thus direct measurement and identification are often impossible. ESR is the only technique which has the potential for direct detection of radicals, but in biological systems even these must be trapped by a spin-trapping agent. Thus, most investigations involve recognition of 'indicators' of the presence of radicals in vivo or 'Footprints' of radical-mediated damage.

In this volume we have assembled all the methods used in our respective laboratories emphasizing the methodologies and techniques that are considered to be the most relevant, appropriate and least subject to error, whilst presenting critical assessments of all the approaches described. Where major specialized, high-cost instrumentation is required the text emphasizes the interpretation of the data obtained, the advantages and disadvantages of the technique and the chemical basis of the approach, rather than the specific details of how to operate the relevant instrument.

We have focussed the experimental approaches towards the measurement of radicals and radical-mediated damage in chemical systems, in cells and in tissues under the following headings:

(a) Footprints of DNA damage
- oxidized bases such as thymine glycol, 8-hydroxydeoxyguanosine,
- adducts of DNA bases,
- thiobarbituric acid-reactive materials (non-specific).

(b) Footprints of protein damage
- detection of oxidized amino acid side-chains by amino acid analysis,
- fluorescence detection of oxidized components,
- measurement of carbonyl-containing products,
- assay of oxidized protein thiol groups in membranes,
- assessment of secondary structural modifications.

(c) Footprints of lipid peroxidation
- detection of malonyldialdehyde by direct assay (HPLC),
- detection of other secondary metabolites of oxidation of polyunsaturated fatty acids such as 4-hydroxynonenal and hexanal by HPLC,
- measurement of lipid hydroperoxides,
- detection of volatile hydrocarbons (ethane, pentane, ethylene),
- detection of oxidation products of cholesterol,
- detection of thiobarbituric acid-reactive material,
- fluorescence detection of aminoiminopropene cross-links,
- chemiluminescence.

(d) Footprints of antioxidant consumption
- decreased GSH, appearance of GSSG,
- tocopherol levels,
- carotene levels,
- selenium concentration,
- antioxidant enzymes.

(e) Footprints via indirect radical assays
- ESR spin-trapping,
- aromatic hydroxylation,
- deoxyribose assay,
- markers of superoxide radical production.

(f) Footprints via the availability of transition metal complexes
- bleomycin assay for available iron,
- ferrozine assay for non-haem iron,
- fluorescent detection of haem iron,
- detection of ferryl species.

There is some confusion in the literature over the terms 'oxidation' and 'oxidative-damage' and the concept of 'radical-damage'. These are in no sense synonymous: radical attack may involve oxidation, reduction or neither, whereas 'oxidative-damage' need not proceed *via* a radical mechanism. Possibly, the confusion arises because oxygen is a major primary source of reactive radicals and the biological conversion of oxygen into water can proceed *via* radical formation (although, generally, no 'damage' as such is involved). However, although hydroxyl radicals can be thought of as being 'oxidizing' species, superoxide ions, for example are good reducing agents, especially in the narrow definitions which equate electron-donors with reducing agents and electron-acceptors with oxidizing agents.

Thus, in a book about radicals, it might have been best to try to avoid the use of terms such as 'oxidative-damage'. However, this is *so* widely used in the bio-medical field that it is difficult to avoid. We therefore use such terms, but with the reservations indicated above.

The assistance of several postdoctoral and doctoral fellows must be acknowledged; in particular, Dr. Patrick McCarthy for his major contribution to the chapter on techniques for measuring lipid peroxidation; Dr. Parves Haris for his assistance and expertise in the preparation of the section on Fourier Transform Infra Red Spectroscopy, Mr. George Paganga for the section on the preparation of low density lipoproteins and Dr. Paul Eggleton for the preparation of the section on the nitro blue tetrazolium assay.

CR-E and ATD are grateful to the Association of International Cancer Research, The Wellcome Trust, and C.R.-E. to The British Heart Foundation, The British Technology Group, and Bioxytech – Paris for their generous support of the associated research which involves many of the techniques described in this book.

The authors thank Wendy Webster and Sylvia Williams for expert secretarial assistance.

Catherine Rice-Evans
Anthony T. Diplock
Martyn C.R. Symons

Contents

Chapter 4.   *Transition metal complexes as sources of radicals* . .   *101*

Introduction to free radicals

1.1. Introduction

This is the introductory chapter which sets the stage for those that follow. We start by attempting the definition of the term 'free radical' together with a very brief historical background. This is followed by brief overviews of some of the types of radical that might be encountered in biological systems.

1.1.1. What is a free radical?

Definitions are odious, difficult to formulate and riddled with loopholes! A broad definition of a free radical is that it is a molecule or ion containing an unpaired electron. The significance of this can be gauged by considering reaction [1] in Scheme 1.1. Here A$\cdot$ and B$\cdot$ are radicals, the dots signifying unpaired electrons. These have paired in the bond between the groups in A-B, which is a normal molecule or ion. (It is treated herein as neutral purely for convenience.) Whilst most radicals are reactive and undergo dimerization, or other reactions, in which the unpaired electrons have become paired, some are relatively stable and have long life-times. These include nitroxide radicals, $R_2\dot{N}O$, and a range of radicals and radical ions in which the unpaired electron is so delocalized that it is unwilling to participate in a localized electron-pair bond. Thus, although to many people, the unique reactivity of a radical lies in the desire of its characteristic unpaired electron to participate in covalent (electron-pair) bonding, there are many exceptions. Other reactions of great importance which help to characterize radicals and which may accompany reactions of type [1] include [2] and [3] (Scheme 1.1) which usually generate new radicals which will ultimately be 'destroyed' via reactions of the type in Scheme 1.1. Sometimes one of these new species has a long life-time and becomes readily detectable, especially by ESR spectroscopy (see under 'spin-trapping', below). Alternatively, [2] or [3] may repeat, so that one initial radical triggers or initiates a series of radical steps before reaction [1] occurs. This is called a *chain reaction* and is highly characteristic of many radicals under the right conditions. The very important *autoxidation* reaction, discussed elsewhere (Chapter 2), is a key biological reaction which owes its importance to its *chain* nature.

1.1.2. Other systems containing unpaired electrons

Normally the term 'radical' (or 'free radical') is confined to molecules or ions with one unpaired electron (called *doublet* states because the electron has two magnetic quantum numbers $\pm 1/2$). In the non-metal field the most common paramagnetic species other than radicals are those with two unpaired electrons, called *triplet* states (magnetic quantum numbers $0, \pm 1$). Quite the most important triplet-state molecule is dioxygen and it is a great pity that ESR spectroscopy can, for various reasons, only be used to detect O_2 in the gas-phase or certain crystalline solids. Other important triplet-states are sometimes obtained on photo-excitation of ordinary (singlet-state) molecules or ions, and these have reactions in some ways typical of di-radicals (i.e.,

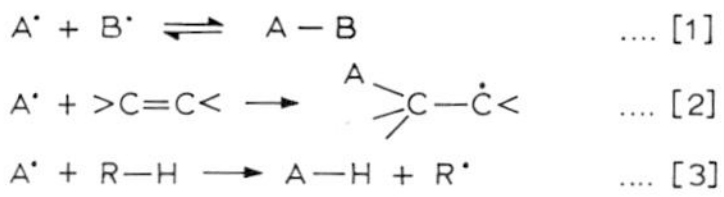

$$A^{\cdot} + B^{\cdot} \rightleftharpoons A-B \qquad \dots [1]$$
$$A^{\cdot} + >C=C< \longrightarrow {}^{A}_{>}C-\dot{C}< \qquad \dots [2]$$
$$A^{\cdot} + R-H \longrightarrow A-H + R^{\cdot} \qquad \dots [3]$$

Scheme 1.1. Some characteristic reactions of radicals. $A^{\cdot}$ and $B^{\cdot}$ are any two types of reactive radical. $A^{\cdot} = B^{\cdot}$ reaction $[1]_+$ is a dimerisation. The reverse $[1]_-$ is bond homolysis and may be induced thermally or photochemically. Reaction [2] represents addition to an alkene derivative. Reaction [3], hydrogen atom transfer, is one of the most important displacement reactions of radicals.

species with two *radical* centres in the molecule). Carbenes and nitrenes are examples of highly reactive ground-state triplet species but, apart from O_2, we do not discuss these species herein.

The other important classes of molecules/ions containing more than one unpaired electron are transition-metal and lanthanide ion-complexes. In general (although not always) these do not exhibit the reactions of Scheme 1.1 and hence are not usually classed as radicals. Often they are very stable, examples being high-spin Mn(II) and Ni(II); the major reaction linking them with radicals is that of electron transfer. (Transition-metal complexes can have up to $5(d)$ unpaired

Scheme 1.2. Generalized electron-transfer reactions
(involving molecules D (electron-donor) and E (electron-acceptor)
(a) $D \rightarrow D^{\cdot +} + e^-$ [4]
                                       (radical cation)
                                       $E + e^- \rightarrow E^{\cdot -}$ [5]
                                       (radical-anion)
                                       $D + E \rightleftharpoons D^{\cdot +} + E^{\cdot -}$ [6]

Reactions [4] and [5] only occur under exceptional conditions. Reaction [6] is a typical electron-transfer reaction.
(Note that if D and E are *Not* radicals, the products $D^{\cdot +}$ and $E^{\cdot -}$ *must* be). These are generally called radical-cations and radical anions, respectively. An important 'intermediate' in some cases is the 'charge-transfer' complex ($D^{\delta +} \ldots E^{\delta -}$)

(b) *Some examples:*
  $RS^- + RNO_2 \rightleftharpoons RS^{\cdot} + RNO_2^{\cdot -}$
  $Na^{\cdot} + PhNO_2 \rightarrow Na^+ + PhNO_2^{\cdot -}$
  $Co(II) + O_2 \rightleftharpoons Co(III) + O_2^{\cdot -}$

(c) Excited states as electron-donors and -acceptors:

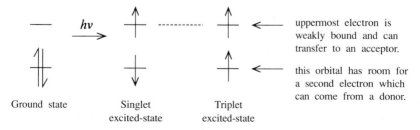

uppermost electron is weakly bound and can transfer to an acceptor.

this orbital has room for a second electron which can come from a donor.

Ground state   Singlet   Triplet
                      excited-state   excited-state

Scheme 1.2. Generalized electron-transfer reactions.

electrons whilst lanthanide complexes can have up to $7(f)$ unpaired electrons.)

1.1.3. Electron-transfer reactions

Molecules, ions or radicals can participate in electron-transfer reactions (Scheme 1.2). These are unique processes in that they involve movement of a fundamental particle (e^-), rather than atom or group transfer. (In this sense, they resemble proton transfers.) If a singlet-state molecule or ion gains or loses one electron, a radical must be formed (Scheme 1.2). (Molecules with all electrons paired have singlet spin states.) Sources of electrons include negative electrodes, anions or molecules with low ionization-potentials, certain photo-excited (often triplet-state, $S = 1$) species, ionizing radiation and electron-rich transition-metal complexes. Good electron-acceptors include positive electrodes, cations or molecules with high electron-affinities, certain excited states, ionizing radiation and electron-poor transition-metal complexes. (The electron donating and accepting properties of excited singlet or triplet states is explained in Scheme 1.2(c).)

1.2. Oxygen-derived free radicals

1.2.1. Dioxygen in triplet and singlet states

A major source of radicals in biological systems is dioxygen (O_2). This very reactive molecule, whilst being essential to the life of higher organisms is, nevertheless, very dangerous in excess or, for example, in ischaemia-reperfusion situations. Dioxygen is a ground-state triplet ($S = 1$), the two unpaired electrons being accommodated, formally, in the degenerate pair of antibonding π^*–orbitals, π^1_x, π^1_y. It is noteworthy that, because of strong coupling to rotational levels, the ESR spectrum for O_2 as a low-pressure gas comprises sets of many narrow lines spread over a wide field range. Unfortunately, these are so extensively broadened for O_2 in solution that no resonance is detectable.

However, oxygen is a source of internal fluctuating magnetic fields, which may broaden the ESR features of other radicals. There are several low-lying excited states for dioxygen, probably the most important being the $^1\Delta$ state in which the two electrons are, formally, paired in one of the π^*-orbitals, leaving the other vacant. The $(\pi_x)^2$, $(\pi_y)^0$ description is not strictly correct for the gas-phase molecule, but we suggest that it is suitable for $^1\Delta O_2$ in aqueous solution. This description stems from our knowledge of the asymmetric solvation of $O_2{}^{\cdot-}$, in which hydrogen bonding is directed towards the filled π^*-orbital only (cf. Fig. 1.1). Some elements of this solvation might also occur for aqueous solutions of the $^1\Delta$ form of dioxygen.

Dioxygen in its ground (triplet) state reacts as a rather stable diradical. One of its most important radical reactions is [4] in which radicals R$^{\cdot}$ are converted into peroxyl radicals R-OO$^{\cdot}$ which generally have quite different reactivities from those of the parent R$^{\cdot}$ species. This reaction is, in general, reversible, and stable radicals such as nitroxides fail to undergo reaction [1.1] to any measurable extent.

$$R^{\cdot} + {}^3O_2 \rightarrow ROO^{\cdot} \qquad [1.1]$$

In contrast, $^1\Delta O_2$ oxygen, often just referred to as 'singlet oxygen', reacts as an electrophile, via the 'empty' π^* orbital. Thus its reactions

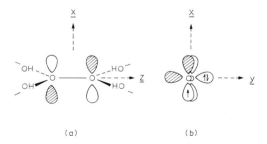

(a)   (b)

Fig. 1.1. Superoxide ($O_2{}^{\cdot-}$) ion: structure and solvation. (a) View along y showing the orbital of the unpaired electron and four water molecules forming hydrogen bonds in the $y - z$ plane. (b) View along z, showing the filled π_y^* orbital and the half-filled π_x^* orbital.

and reactivity are rather different. Of course, being an excited state, it has a limited life-time, and in the absence of reaction it falls to the ground state with light emission (in the 1270 nm region for spontaneous, unimolecular emission (Kanofsky 1983, 1984)). In the absence of collisions (e.g., in the upper atmosphere) its life-time is probably very long (ca. 1 h) (Badger et al., 1965), but in solution, the quantum efficiency of light emission falls drastically, even though its rate is greatly increased (Scurlock and Ogilby, 1987). Probably solvent complexes are formed which facilitate both radiationless decay and light emission.

Singlet oxygen is formed from excited-states of various sensitisers such as acridine. It is thought to be of importance in various biological contexts, one important example being 'photodynamic cancer therapy'. We stress, however, that it is normal triplet oxygen that has radical characteristics, not singlet oxygen. The importance of singlet oxygen is discussed in more depth in Section 3.4.

Returning to oxygen in its ground-state, it is important to mention that this is also toxic when its concentration is not properly controlled, and high concentrations of oxygen can lead to cell death. Oxygen toxicity is associated with reductions in triphosphate levels, alterations in the electron transport chain, depletion of glutathione and various other cellular changes. It seems probable that one cause of oxygen toxicity is the generation of oxygen centred radicals, $O_2^{\cdot-}$ and $\cdot OH$, and of hydrogen peroxide. This is discussed further in Chapter 2.

1.2.2. Superoxide ions

The *superoxide ion* has its unpaired electron in a π^* orbital (Fig. 1.1). There is no distinction between π^*_x and π^*_y in the absence of solvation but one of us has presented ESR evidence for strong hydrogen-bonding primarily on one plane, as is also indicated in Fig. 1.1. (Symons et al., 1980; Symons and Stephenson, 1981). This favours the filled, negative, π^* orbital, forcing the unpaired electron into the 'unsolvated' orbital.

No ESR signal has ever been observed for $O_2^{\cdot-}$ in liquid water.

This is for a variety of reasons (see San Filippo et al., 1976; Boon et al., 1988) and is most unfortunate (see Section 3.1). Indeed, it is a curious coincidence that all the small, fundamental, free radicals that are involved in the initiation of many biological processes are, for various reasons, undetectable by ESR spectroscopy in fluid aqueous solution (these include O_2, $O_2^{\cdot-}$, $HO_2^{\cdot}$, $^{\cdot}OH$ and $^{\cdot}NO$).

Crystalline salts of the superoxide ion, such as KO_2, are commercially available, and provide a ready source of $O_2^{\cdot-}$ in solution. Alternatively, it can be formed by electrolysis, by ionization of $HO_2^{\cdot}$, readily formed from H_2O_2, or be electron-addition to O_2. The superoxide ion $O_2^{\cdot-}$, is a weak base, as well as being a radical. Its basicity and its nucleophilicity are a strong function of solvation, as for all anionic bases, increasing enormously as the extent of hydrogen-bonding solvation is reduced. This has two important implications: (1) on initial formation from O_2, $O_2^{\cdot-}$ will be a far stronger base than it is once fully hydrated; and (2) when $O_2^{\cdot-}$ is formed from O_2 in a non-aqueous environment, such as a lipid membrane, or a hydrophobic pocket of a protein, it will be a strong base, and this could be its overriding mode of reaction in view of its low reactivity as a radical (Niehaus, 1978; San Filippo et al., 1976).

This is nicely illustrated in work on the generation of $O_2^{\cdot-}$ from O_2 in dry dimethylformamide, which gave a species whose ESR spectrum agrees closely with that expected for an $RO_2^{\cdot}$ (peroxyl) radical, not $O_2^{\cdot-}$, which would probably remain undetected in this solvent (Boon et al., 1988). We suggest that reaction [1.2], comprising *nucleophilic* addition to the carbonyl group, is responsible.

$$O_2^{\cdot-} \quad + \quad \underset{Me_2N}{\overset{H}{>}}C=O \quad \rightleftharpoons \quad \underset{Me_2N}{\overset{H}{>}}C\overset{OO^{\cdot}}{\underset{O^-}{<}} \qquad [1.2]$$

Addition of water readily reverses this reaction. However, for esters, the reaction proceeds as in reaction [1.3], and it has been suggested that this could be an important reaction for $O_2^{\cdot-}$ generated in lipid membranes. Both the intermediate and the acylperoxyl radical should be more reactive as *radicals* than $O_2^{\cdot-}$.

$$O_2^{\bullet-} + \begin{array}{c} R \\ \diagdown \\ R'O \end{array} C=O \rightleftharpoons \begin{array}{c} R \\ \diagdown \\ R'O \end{array} C \begin{array}{c} OO^{\bullet} \\ \diagup \\ \diagdown \\ O^- \end{array} \longrightarrow \begin{array}{c} OO^{\bullet} \\ \diagup \\ RC \\ \parallel \\ O \end{array} + R'O^- \qquad [1.3]$$

These considerations are important because of the high solubility of oxygen in non-aqueous solvents relative to that in water. Of course, since $O_2^{\bullet-}$ is greatly stabilized by water it will be a far better *electron-donor* in aprotic solvents than it is in water. Thus in reaction [1.4] where A is,

$$O_2^{\bullet-} + A \rightleftharpoons O_2 + A^{\bullet-} \qquad [1.4]$$

say, an aromatic compound, the solvation of $O_2^{\bullet-}$ will be far greater in water than that of $A^{\bullet-}$. Hence equilibrium [1.4] might well favour $O_2^{\bullet-}$ in water, but $A^{\bullet-}$ in a lipid environment.

$RO_2^{\bullet}$ and $HO_2^{\bullet}$ are more reactive than $O_2^{\bullet-}$ as radicals, and can extract 'active' hydrogen as, for example, in reaction [1.5].

$$RO_2^{\bullet} + R'CH_2CH=CHR'' \rightarrow ROOH + R'\dot{C}H\overline{CHCH}\dot{C}HR'' \quad [1.5]$$
(an allyl radical)

1.2.3. Hydroperoxyl and peroxyl radicals

The radical $HO_2^{\bullet}$, formed reversibly from $O_2^{\bullet-}$ in acidic solutions ($pK_a = 4.7$) is similar in reactivity to $RO_2^{\bullet}$ radicals, formed by oxygen addition to alkyl and related radicals, or by H-atom loss from ROOH molecules. As radicals, they are more reactive than $O_2^{\bullet-}$ ions and can extract 'active' hydrogen, as in reaction [1.5] to give allyl radicals. The resulting hydroperoxide, like H_2O_2, is quite reactive, as is the allyl radical which, formally, has ca. 50% spin-density on two almost equivalent carbon atoms, as indicated. Reaction [1.5] is of major importance in certain membrane reactions, since it is one of the stages in the autoxidation of unsaturated lipid groups (see Chapter 2). The total reaction comprises [1.5] followed by [1.6] which, in turn is followed by [1.5]. These two steps constitute a chain reaction which may continue

$$R' \overset{\bullet}{C}H \overline{CH} \overset{\bullet}{C}HR'' + O_2 \longrightarrow R' - CH - CH = CHR''$$

$$| \atop O \searrow O^\bullet \qquad [1.6]$$

$$(RO_2^\bullet)$$

until the supply of oxygen runs out. The key factor is, of course, that one active radical formed in the membrane may lead to many damaged lipid groups. Such chain reactions require a fine balance of reactivities and are relatively rare in biological systems. However, when they do occur they have an importance which is far greater than that of non-chain reactions.

We have now seen two very important roles for dioxygen, namely, electron addition and addition to carbon-centred (and other) radicals to give $RO_2^\bullet$ (peroxyl) radicals. Both are of great importance in biology, although, generally, biological reactions avoid both, unless the products ($O_2^{\bullet -}$ and $RO_2^\bullet$) are needed in some specific process (phagocytosis is a possible example). Thus enzymes involved in the conversion of dioxygen into water react either via the direct addition of two electrons, or via the addition of one, followed rapidly by another, within an enzyme pocket. This gives H_2O_2 (or HO_2^-) and even these intermediates can be avoided if more electrons are available. Then the overall '4-electron' reaction, giving $2H_2O$, occurs 'safely' within the reaction pocket. Possibly one danger is to overwhelm such enzyme systems such that they are unable to return to their resting forms before another substrate molecule (such as oxygen) is taken up. Hence the normal 2- or 4-electron stage is unable to go to completion. An example seems to be the formation of $O_2^{\bullet -}$ and H_2O_2 by xanthine oxidase. This enzyme has the ability to handle multi-electron reactions so $O_2^{\bullet -}$ formation is probably anomalous. If active radicals *are* formed, there are specific enzymes which readily scavenge and decompose them. The superoxide dismutases are good examples, in the catalysis of reaction [1.7].

$$2O_2^{\bullet -} + 2H_2O \rightleftharpoons H_2O_2 + O_2 + 2OH^- \qquad [1.7]$$

1.2.4. The hydroxyl radical

This is the most reactive of the oxygen centred radicals, but its high reactivity does not mean that it is the most important (see Chapter 3.2). Hydroxyl radicals are formed from water by electron-loss ($H_2O^{\cdot+}$ is a strong acid and rapidly loses a proton) but this is normally only achieved by ionizing radiation. The most important mode of generation is from H_2O_2, either by bond homolysis (UV-light) or by electron capture [1.8]. It is generally agreed that certain transition-metal aquo-ions can achieve this, but complexed ions may react by alternative routes (see Chapter 4).

$$H_2O_2 + \text{`e}^-\text{'} \rightarrow OH^- + {}^\cdot OH \qquad [1.8]$$

The importance of $^\cdot OH$ radical reactions in tissues is that they are *so* reactive that they tend to react with the first bio-molecule that they encounter. They react mainly by H-atom extraction or by addition to double bonds.

$$RH + {}^\cdot OH \rightarrow R^\cdot + H_2O \qquad [1.9]$$

$$\diagup\!\!\!\!\!>\!C = C\!<\diagdown\!\!\!\!\! + {}^\cdot OH \longrightarrow \underset{\diagup\,|}{\overset{HO\diagdown}{C}}\!-\!\dot{C}\!< \qquad [1.10]$$

Thus damage is widespread but indiscriminate. They will not travel far, and reaction schemes in which they are generated in one part of a cell and thought to react in another part are likely to be incorrect. However, systems which generate $^\cdot OH$ close to a given substrate, such as DNA, will be effective in inducing damage. An example is bleomycin, which binds strongly to DNA and has a high affinity for ferrous ions. These can react with H_2O_2 to give $^\cdot OH$, which then attacks DNA with high probability.

The pK_a of $^\cdot OH$ is close to that of water, so O^- ions are unlikely to be of biological importance. The $^\cdot OH$ radical has a high electron-affinity and its reactions are sometimes described as being 'electrophil-ic' in nature. Thus it prefers to react at electron-rich sites in molecules, although, in fact, it is not very selective.

We conclude that, although the ·OH radical is the most reactive of the 'oxygen radicals', it may not be as dangerous as, say, $O_2^{·-}$, be-cause of the speed and indiscriminate nature of its reactions. Only when it is generated very close to its target is it expected to be impor-tant.

1.3. Nitric oxide

It is only relatively recently that the biological importance of nitric oxide (·NO) has been appreciated (Palmer et al., 1987; Moncada et al., 1988). This radical has a structure similar to that for $O_2^{·-}$, except that it has 2 e^- less, i.e., it is isoelectronic with O_2^+, just as NO^- is a triplet-state species iso-electronic with O_2. It has only a minor ten-dency to dimerise, and it is not clear to what extent its reactivity in biological reactions is linked to its radical character. It is very readily oxidised to ·NO_2 whose possible importance in biological reactions seems to be unknown. These two paramagnetic oxides are closely re-lated to NO_2^- and NO_3^- ions. For example, NO_2 disproportionates to give nitrite and nitrate in acidic solution, but NO_2^- in acid gives NO_3^- + ·NO [1.11].

$$3NO_2^- + 2H^+ \rightleftharpoons NO_3^- + 2NO^· + H_2O \qquad [1.11]$$

Thus nitrite ions are a direct source of NO in acidic media. The other important reaction is that with oxygen [1.12], which is quite rapid and effectively irreversible. This means that NO· will only reach high con-centrations in the absence of oxygen.

$$2NO^· + O_2 \rightarrow 2NO_2 \qquad [1.12]$$

The biochemical generation of NO is thought to occur from the –NH–

$C(NH_2)_2{}^+$ unit of arginine, via the formation of hydroxylamine. It can be characterized by ESR spectroscopy using deoxy-myoglobin or -haemoglobin with which it reacts to give nitrosyl derivatives having well-defined spectra. Its chief biological role seems to be as a vasorelaxant (Moncada et al., 1988).

1.4. Carbon and carbon-oxygen centred radicals

1.4.1. Carbon-centred radicals

Small alkyl radicals such as methyl ($^{\cdot}CH_3$) do not seem to play an important role in biology. (Except that it might be significant that methyl radicals are readily detected by ESR spectroscopy in various flints (cherts) (Griffiths et al., 1982). One wonders whether, in a methane-rich atmosphere, methyl radicals might be of some biological importance.)

However, allyl-type radicals are thought to be important in the autoxidation of membrane lipids etc. (Chapter 5). These are more stable because the unpaired electron is delocalized on the two outer carbon atoms.

1.4.2. Carbon-oxygen centred radicals

Delocalisation onto oxygen stabilizes radicals considerably. An important example is the ascorbate radical (Scheme 1.3) formed by electron-loss from the ascorbate anion, or electron-capture by dehydroascorbate. This is remarkably stable, and is characterized by an ESR doublet (1.7 G) which is quite distinctive. Because of the high sensitivity of ESR spectroscopy, and the fact that opaque samples can be used, ascorbate radical intermediates have been widely studied (Liu et al., 1988a). The most probable structure is shown in Scheme 1.3 but this is still a matter of some controversy (Liu et al., 1988a). A key factor in the formation of ascorbate radicals is that ascorbate anions

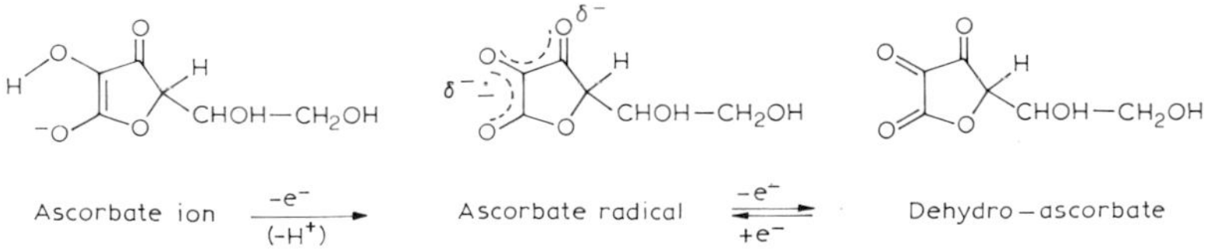

Scheme 1.3. Involvement of ascorbate in electron-transfer reactions.

react with a wide range of radicals at near diffusion-controlled rates, whereas ascorbate radicals react slowly. However, in electron-transfer reactions, the ascorbate radical is by far the most reactive. Analogies between the ascorbate and nitroxide systems are discussed in the next section (1.5).

Many radical intermediates have structures related to the relatively stable benzosemiquinone radical (Scheme 1.4). Some examples of biologically important derivatives are given in Fig. 1.2. Quinones are thought to be involved in redox cycling, the key step being electron-donation from the semiquinone anions to oxygen to give superoxide radicals. Their mediation seems to be important in increasing overall reaction rates. They may also participate in electron transport chains, for example, mitochondria. Another important reaction of certain

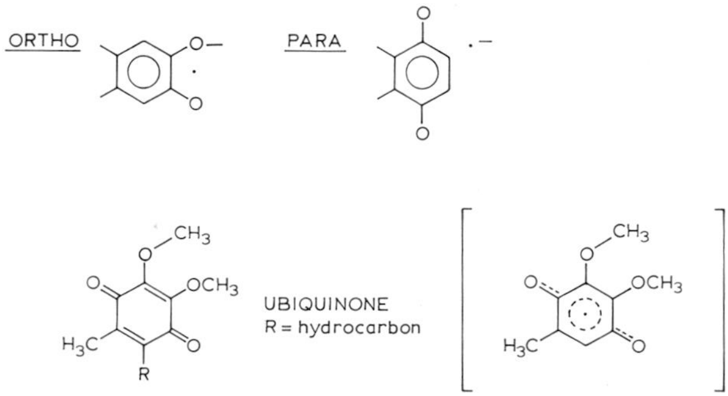

Scheme 1.4. Structure of Bio 'quinones'.

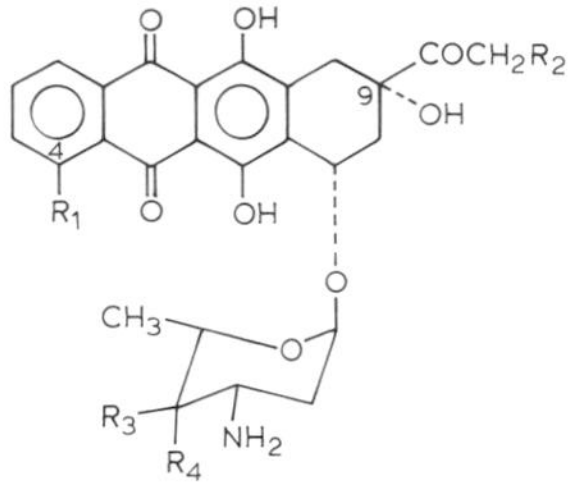

Anthracycline	R_1	R_2	R_3	R_4
Adriamycin	OCH_3	OH	H	OH
4' Epiadriamycin	OCH_3	OH	OH	H
4' Deoxyadriamycin	OCH_3	OH	H	H
Daunorubicin	OCH_3	H	H	OH
4' Demethoxydaunorubicin	H	H	H	OH
Carminomycin	OH	H	H	OH

Fig. 1.2. The structures of the anthracycline analogues.

quinones is to accept electrons (or H-atoms) from RSH derivatives (Grant et al., 1988).

1.4.3. Melanin and tyrosine radicals

Very few radicals exist in tissues 'at rest'. Two important exceptions are melanin radicals and tyrosine radicals. The former exist in low concentrations in samples of melanin, which is a high polymer made up of units which include potential semiquinone units. It is reasonable to expect some *ortho*-semiquinone formation from such structures. The radicals can be thought of as occluded, being sterically protected by surrounding polymer. So far as we know, they have no special chemical significance.

Similarly, it is possible that tyrosine radicals are stabilized in part by being in a protective environment. Phenoxyl radicals in general are quite reactive and it is surprising that this particular radical should be readily detected in various biological systems (Ehrenberg and

Reichard, 1972; Barry and Babcock, 1987). It is probably formed by electron-loss from the tyrosine side-group, followed by proton loss, since it is found in redox enzymes such as ribonucleotide reductase, which catalyses the conversion of ribonucleotides into deoxyribonucleotides. In this enzyme the tyrosyl is said to be 'stabilized' by an adjacent binuclear iron unit, but the mode of stabilization is not clear, and the ESR and ENDOR spectra of the radical are not significantly perturbed, as would have been expected if any direct bonding were involved.

1.5. Nitroxide radicals

Although nitroxide radicals do not seem to exist in naturally biological systems, they are relatively non-toxic, and survive as such for considerable periods after administration. This property makes them important as 'spin-labels' in biological systems, ESR spectroscopy being an ideal technique for studying their behaviour. However, in these studies their radical nature is not significant chemically: it is the presence of an unpaired electron, and the remarkable stability of these species that is utilized. However, nitroxides can act as electron donors and acceptors, and we compare them here with ascorbate radicals (Section 1.4).

The reason for the stability of nitroxides is 2-fold: on the one hand, the unpaired electron is almost equally shared by nitrogen and oxygen (in a π^*-orbital). In the dimers [1.13] these electrons are forced into the N-N σ-bond, and the delocalization has to be overcome to do this.

$$
\begin{array}{c}
O \\
R \triangleright N \\
R
\end{array}
\quad \longleftarrow\!\!\!\|\!\!\!\longrightarrow \quad
\begin{array}{c}
O \\
N \triangleleft R \\
R
\end{array}
\qquad [1.13]
$$

Hence the bonds are long and weak, and readily break. Also, all stable nitroxides, such as $(Me_3C)_2 NO^{\bullet}$, have very bulky R-groups which sterically inhibit dimerisation. The other important factor is that disproportionation [1.14] is not favoured.

$$2R_2NO^{\bullet} \rightleftharpoons R_2NO^{+} + R_2NO^{-} \qquad [1.14]$$

This contrasts with the ascorbate system in which two ascorbate radicals give ascorbate + dehydroascorbate.

Nevertheless, the ionization potential for $R_2\overset{\bullet}{N}O$ radicals is quite low so they can act as e^{-}-donors, and they can also act as e^{-}-acceptors, especially in the presence of proton donors which react to give R_2N-OH molecules. (NB: R_2NO^{+} cations are isoelectronic with ketones and hence should be relatively stable. They can in principle add water to give $R_2N(OH)_2^{+}$ or $R_2N(OH)O$ derivatives.)

It is important that nitroxides can oxidize ascorbate anions and thiols, by one-electron transfer processes. Indeed, when ascorbate is added to a nitroxide solution the ESR spectrum of the nitroxide is largely replaced by that of ascorbate (Liu et al., 1988b).

The other, very important role played by nitroxides in radical chemistry is via the use of spin-traps which are converted into nitroxides on reaction with reactive radicals (e.g., reactions [1.15] and [1.16]).

$$RNO + X^{\bullet} \longrightarrow \overset{X}{\underset{R}{\diagdown}}NO^{\bullet} \qquad [1.15]$$

$$R-CH=N-R' + X^{\bullet} \longrightarrow RCH(X)-N-R' \qquad [1.16]$$
$$\underset{O}{|} \qquad\qquad\qquad\qquad \underset{O^{\bullet}}{|}$$

In this way radical intermediates which cannot be directly detected in fluid systems are converted into stable radicals which accumulate in the system and can be studied by ESR spectroscopy. This process is discussed in Chapter 3 (Section 3.2).

1.6. Sulphur-centred radicals

Radicals formed from thiols and disulphides may play important roles in cellular function and malfunction, the RS· radical, and RS-SR⁻ radical anions being particularly implicated. The pK_a values for many thiols, in the region of 9 to 10, are such that there are low, but possibly significant concentrations of RS⁻ anions at pH.7. Thus RSH may act as an H-atom donor or, via RS⁻, as an electron donor. As with other radicals of this type, RS· radicals have not been detected by ESR spectroscopy in aqueous solution, nor do they have intense absorption bands in the near-UV. However, they have a high affinity for RSH and RS⁻ (Reactions [1.17]–[1.18]) which act as 'spin-traps' provided they are present in fairly high concentrations.

$$RS\text{·} + RS^- \rightleftharpoons RS \dot{-} SR^- \qquad [1.17]$$

$$RS\text{·} + RSH \rightleftharpoons RS \dot{-} SHR \rightleftharpoons RS \dot{-} SR^- + (H^+) \qquad [1.18]$$

The radical anions RS-SR ⁻ are remarkably stable, and act primarily as electron donors. They are readily detectable in aqueous solution, both optically (an intense band at ca. 410 nm) and by ESR spectroscopy. They can, of course, be derived from RS-SR compounds by direct electron-addition.

It is interesting to note the marked differences between oxygen and sulphur in this context. Thus RS-H compounds are good H-atom donors, but water and alcohols are not. RS⁻ anions are good electron donors, and are readily formed in dilute alkaline solution, in contrast with alkoxide ions (RO⁻). RS· radicals add to RSH or RS⁻ to give RS-SR ⁻. In contrast, RO-OR ⁻ ions are not formed, and RO· radicals are powerful radical agents, adding readily to double bonds, and extracting hydrogen atoms from many types of molecule.

1.7. Conclusions

A summary of various classes of radicals has been given above. Obviously many radicals have not been directly specified but, hopefully, those of biological importance have largely been covered. In the following chapters many points raised herein are elaborated. In Chapter 2, we cover different ways in which radicals are generated, and ways in which biological systems protect themselves from radical induced damage. In Chapter 3, methods of direct and indirect detection of radical intermediates are outlined.

Chapter 4 is concerned with the involvement of transition-metal complexes and metallo proteins in radical generation, elaborating on these important reactions, and mentioning alternative ways of formulating the overall reactions involved. In Chapters 5–8, attention is turned to specific biological systems: 5 and 6 being primarily concerned with lipid damage and antioxidants, respectively, 7 with proteins, and 8 with DNA and base damage.

CHAPTER 2

Mechanisms of radical production

2.1. Introduction

Radicals are formed by a wide variety of reactions, and a few are so stable that there is no need for specific generation. As stressed in Chapter 1, these include dioxygen, in its ground-state triplet form, nitric oxide ($^\cdot$NO) and nitrogen dioxide ($^\cdot$NO$_2$), superoxide ion (O$_2^{\cdot-}$) in its salts (as KO$_2$), a range of nitroxide radicals (R$_2$ṄO), and various complex organic radicals (Scheme 2.1) such as perinaphthyl radical (I). Stable radicals occurring in tissues include the two types of melanin radical (II, III) and the tyrosyl radical (IV) which occurs naturally in certain enzymes.

We rather arbitrarily consider radical production in terms of physical, chemical and biological sources.

2.2. Physical generation of radicals

2.2.1. Ultraviolet light

Photolysis of light sensitive compounds (A-B) with light in the 250–400 nm range generally results in bond homolysis [2.1], with:

$$A\text{-}B + h\nu \rightarrow A^\cdot + B^\cdot \qquad [2.1]$$

the formation of two radical centres. Higher energy photons may give photo-ionization [2.2]

$$A\text{-}B + h\nu \rightarrow AB^{\cdot+} + e^- \qquad [2.2]$$

and certain molecules or ions react in this way even with near UV light. A third mechanism involves homolytic reactions of excited singlet or (more often) triplet states (AB*), again generating two radicals [2.3 and 2.4]:

$$AB + hv \rightarrow AB^* \qquad\qquad [2.3]$$
$$AB^* + \text{R-H} \rightarrow {}^\bullet ABH + R^\bullet \qquad\qquad [2.4]$$

Excited states can also act as electron donors, via the electron in the outer orbital [2.5], or electron-acceptors, via the half-filled lower orbital [2.6]

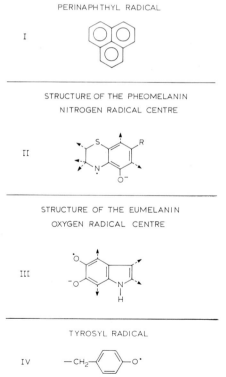

PERINAPHTHYL RADICAL

I

STRUCTURE OF THE PHEOMELANIN
NITROGEN RADICAL CENTRE

II

STRUCTURE OF THE EUMELANIN
OXYGEN RADICAL CENTRE

III

TYROSYL RADICAL

IV $-CH_2-\!\!\bigcirc\!\!-O^\bullet$

Scheme 2.1. Stable complex radicals and stable radicals occurring in tissues.

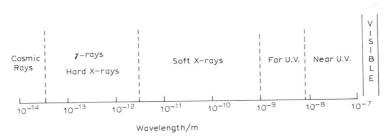

Fig. 2.1. The range of ionizing radiation.

$$AB\,(\uparrow)\,(\uparrow) + A \rightarrow AB^{\bullet+}\,(\uparrow) + A^{\bullet-} \qquad [2.5]$$

$$AB^*\,(\uparrow)\,(\uparrow) + D \rightarrow {}^{\bullet}AB^-\,(\uparrow\downarrow)\,(\uparrow) + D^{\bullet+} \qquad [2.6]$$

When using relatively high power laser light sources, bi-photonic absorption may occur concurrently with normal light absorption. Commonly this results in photo-ionization.

2.2.2. Ionizing radiation

This term includes high-frequency electromagnetic radiation, from UV via X-rays to γ-rays (Fig. 2.1) and also high energy particles such as neutrons, electrons, protons or α-particles (helium nuclei). These may be produced using particle accelerators or from radioactive nuclei. In many cases, initial interaction involves the ejection of an electron (Fig. 2.2). This often has sufficient energy to eject another electron at a later stage, so that tracks or spurs of events are involved rather than single events, as in photo-ionization. Thus, although the physics of the initial interaction is a complicated function of the energy involved, it is reasonable to assume that the most important events involve ejection of high-energy electrons followed by e^--capture. For neutrons, one can again simplify and argue that the most significant mode of energy deposition involves the initial formation of protons, followed by electron ejection.

Although the nature of the primary products seems to be largely governed by simple ionizations, the spatial distribution of these events

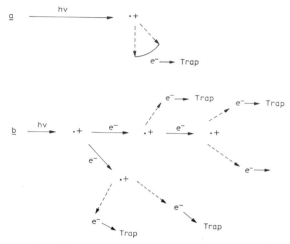

Fig. 2.2. Pictorial view of (**a**) photoionization (UV light) and (**b**) high-energy ionization.

does result in highly heterogeneous initial concentrations, and radical-radical reactions in these regions will, in effect, reduce the initial yield of *available* radicals, more or less, depending on the radiation involved.

Here we confine our attention to simple ionization giving electron-loss centres, and capture of thermalized electrons giving electron-gain centres. These are, of course, generally radical centres, since the parent molecules or ions are generally non-radical species. The complications that arise with multiple damage sites are especially important with protons, α-particles and heavy ions, and are undoubtedly of major biological importance. However, this is beyond the scope of the present work, and we simply focus attention on radiolysis as a source of radicals.

A summary of important events is given in reactions [2.7] – [2.9] where A-B is any molecule or ion system. In a solution containing several components, competition reactions are important, and, for dilute solutions, damage to the solvent usually dominates, since ionizing radiation is generally unselective.

$$AB \qquad \rightarrow AB^{\cdot +} + e^- \qquad\qquad [2.7]$$
$$AB + e^- \qquad \rightarrow AB^{\cdot -} \qquad\qquad [2.8]$$
$$AB^{\cdot +} + e^- \rightarrow AB^* \qquad\qquad [2.9]$$

Reaction [2.9] represents electron return (not necessarily to its initial source). This generally gives an excited molecule (AB*) which may undergo bond homolysis [2.10]:

$$AB^* \rightarrow A^{\cdot} + B^{\cdot} \qquad\qquad [2.10]$$

or may fall to the ground-state without reaction. This is equivalent to a normal photochemical reaction.

These are the basic reactions, but there are a range of subsequent steps which may follow within a few nanoseconds. These include proton loss from the hole-centre (electron-loss centre), dissociative electron capture for the electron-excess centre [2.11] or protonation of this centre. These

$$AB^{\cdot -} \rightarrow A^{\cdot} + B^- \qquad\qquad [2.11]$$

will prevent, or modify, electron return.

In solid-state studies, ESR spectroscopy is the best detection method for studying radical intermediates in radiolysis. It is, however, difficult to apply to liquid-phase studies, and generally, optical methods are favoured. In solid-state work, radicals are trapped (matrix-isolated) and can be studied by any spectroscopic technique at leisure. However, for liquid-phase studies, time-resolved methods are often necessary because the intermediates are usually very short lived. In the technique of pulse radiolysis, short pulses of radiation, followed by pulses of light which explore the UV spectrum, are used. The spectra help to identify the species, but also their kinetic behaviour can be accurately monitored over very short time-scales (from picoseconds to milliseconds). This technique is discussed in Section 3.3.

2.2.3. Water radiolysis

Since most studies have been conducted using dilute aqueous solutions, the substrate radicals are mainly formed by attack of water radicals. However, although these must be of great importance in biological conditions, it must be remembered that not all of the key systems are dilute aqueous solutions. This is true of lipids, membrane-bound proteins and tightly packed chromatin, for example.

After primary ionization [2.12]:

$$H_2O \rightarrow H_2O^{\cdot +} + e^- \qquad [2.12]$$

the $H_2O^{\cdot +}$ cations rapidly lose protons [2.13], and the 'dry' electrons rapidly become solvated.

$$H_2O^{\cdot +} + H_2O \rightarrow {}^{\cdot}OH + H_3O^+ \qquad [2.13]$$

(Solvated electrons are commonly thought to have a solvation shell somewhat similar to that of a normal anion such as Cl^-. Solvation occurs in less than 1 ps and the resulting e^-_{aq} units are much less reactive than 'dry' electrons). Relatively low yields of hydrogen atoms are also present, possibly formed from H_2O homolysis, or from reaction between dry electrons and H_2O (to give $H^{\cdot}$ and OH^-), and also H_2O_2 seems to be formed at a very early stage. However, the main reactive radicals are ${}^{\cdot}OH$ and e^-_{aq}.

2.2.4. Conversion of primary water radicals into secondary radicals

In pulse-radiolysis studies, it has become common practice to convert these primary radicals into secondary radicals, and then to study their reactions with substrates. In general, this is done in order to simplify the system, an example being the conversion of e^-_{aq} into ${}^{\cdot}OH$ radicals (equation [2.14]). This

$$e^-_{aq} + N_2O + H_2O \rightarrow N_2 + {}^{\cdot}OH + OH^- \qquad [2.14]$$

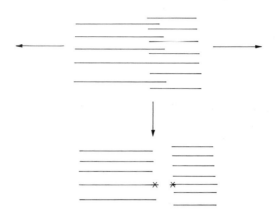

Fig. 2.3. Breakage of a fibre showing how bond-breaking (*) may be a more favourable process than slippage for certain molecules.

is a fairly rapid process, so provided N_2O is in large excess over the substrate, reactions become predominantly those of $\cdot OH$ radicals only. In another type of secondary reaction, $\cdot OH$ radicals are scavenged to give relatively stable products, leaving e^-_{aq} as the major reactant. A further ramification is then to convert hydroxyl radicals into weak electron-donors, such as $\cdot CO_2^-$ [2.15] so that more specific reactions can be induced [2.16]. By converting e^-_{aq}

$$\cdot OH + HCO_2^- \rightarrow H_2O + \cdot CO_2^- \qquad [2.15]$$
$$\cdot CO_2^- + AB \rightarrow CO_2 + \cdot AB^- \qquad [2.16]$$

into $\cdot OH$ radicals with N_2O [2.14], the two major radicals are entirely converted into the electron-donor species $\cdot CO_2^-$.

2.2.5. Mechanical production of radicals

There are two major routes to molecular damage when systems are subjected to mechanical forces. The most obvious is simple bond breaking when linear or cross-linked polymer chains are subjected to shear forces, for example, stretching, cutting, bending, breaking or grinding. Generally, molecules slip past each other, but occasionally polymer molecules undergo homolytic bond breaking when they are held too firmly by their neighbours (Fig. 2.3).

These effects are well known in high polymer chemistry, and bond homolysis has been directly observed by ESR spectroscopy, the broken ends being reasonably well characterized. One remarkable biological example is in the simple act of cutting fingernails, which generates a high yield of sulphur-centred radicals (presumably the weakest links in the chain being S-S bonds) (Chaudra and Symons, 1987).

Another process, associated with flow, is the tribo-electric effect in which electrons are transferred from one molecule to another during flow. This requires contact between donor and acceptor molecules or ions which are then forced apart and retain their charges. As stressed above, these centres are likely to be paramagnetic and should be detected by ESR spectroscopy. This phenomenon is very well established at the electrostatic level for surface charges, but is sometimes hard to distinguish from bond homolysis in several systems (Symons, 1988).

Both could be of some importance in the malfunction of joints and in bone-fracture. In the former, synovial fluid should protect the joint tissue from damage, but if its effect is reduced for some reason, friction under high pressure could cause both types of damage to occur, and any active radicals formed could cause further damage via free radical attack. The same applies to bone fracture. This almost certainly leads to radical formation, but the nature of the radicals and their potential for doing serious damage has yet to be discovered.

2.2.6. Generation of radicals by ultrasound

Sonic waves comprise linear propagation of regions of compression and rarefaction, and are normally quite harmless. The systems normally adjust in-phase or with a lag time depending on the frequency, and the laws of thermodynamics are obeyed (Blandamer, 1973). However, at ultrasonic frequencies, if the power is increased, micro-cavitation may be induced in liquids, caused by thermal fluctuations of the molecules. These micro-bubbles are then subjected to large pressure fluctuations which can give rise to local heating which is sufficient to cause bond homolysis if it is channelled into a few molecules. The resulting radicals may recombine in the bubble or may escape and react elsewhere. ESR spectroscopy with spin-trapping (Reisz et al., 1985) has been used to demonstrate the formation of H$^{\cdot}$ and $^{\cdot}$OH radicals in liquid water by ultrasound.

The power of ultrasonic generators is often in the region of 10^5 Wcm^{-3}. This is sufficient to induce cavitation and generate water or other radicals, and hence the widespread use of ultrasound in medicine should be a matter of possible concern.

2.3. Chemical generation of radicals

The most obvious method for radical generation is to heat molecules until the bonds begin to break.

2.3.1. Thermolysis

If a molecule or ion has one relatively weak bond, precise homolysis [2.17] may occur on heating.

$$A\text{-}B \xrightarrow{\text{heat}} A^{\cdot} + B^{\cdot} \qquad [2.17]$$

An example is the thermolysis of peroxides or of azo-compounds [2.18].

$$R\text{-}N = N\text{-}R \xrightarrow{\text{heat}} 2R^{\cdot} + N_2 \qquad [2.18]$$

However, except in special cases, homolysis is not an important reaction in biology. Perhaps it is the reverse that needs to be considered. Thus, certain radicals do not dimerize or react together, and others dimerize reversibly at ambient temperatures. Examples include $O_2^{\cdot-}$, NO and NO_2, the last forming N_2O_4 reversibly.

2.3.2. Redox processes

Whilst thermolyses have aspects in common with photolyses, redox reactions are closely related to radiation reactions, in that they involve electron-transfer reactions. Electron-transfers, like proton transfers, can be extremely fast processes, and are therefore often key steps in biochemical systems. Probably the best understood example is that of photosynthesis, but there are many others not involving initial light absorption. The basic process is [2.19], but this

$$A + D \rightleftharpoons A^{\cdot-} + D^{\cdot+} \qquad [2.19]$$

comes in many guises. The driving forces are the electron affinities of acceptors and the ionization potentials of donors. One limiting approach is to use electrodes as e^--donors or -acceptors. Another is to use low-valence states of transition metal complexes as e^--donors or high valence states as acceptors. An extremely important example of the former is the Fenton reaction [2.20], Fe^{2+} acting as the donor and H_2O_2 as:

$$Fe^{2+} + H_2O_2 \rightarrow Fe^{3+} + OH^- + \cdot OH \qquad [2.20]$$

the acceptor (see Chapter 4). This type of reaction corresponds to dissociative electron-capture in radiolysis. It converts the relatively stable molecule H_2O_2 into the highly reactive radical $\cdot OH$.

Examples of many of these types of reaction are given throughout this book. We end on a note of caution. The detection of radicals, for example by ESR spectroscopy, does not in itself establish that they are important intermediates in the system under study. The technique is very sensitive, and the radicals detected may be formed by a side reaction of no particular significance. Further study is then required to establish their significance.

2.3.3. Subsequent events

Radicals may 'die' when radical-radical interactions occur, or when they react with transition metal complexes. The former generally comprise 'dimerization' or disproportionation. The latter generally involve charge-transfer.

However, inter-radical conversion may be extremely important, as illustrated throughout, the most important reactions being H-atom transfer or addition to unsaturated compounds. For example, ˙OH radical attack on DNA is dominated by H-transfer to give sugar radicals, and additions to the bases (Chapter 8).

2.4. Normal generation of radicals in biological systems

Free radicals are essential to many normal biological processes. They are involved in cyclo-oxygenase and lipoxygenase action in eicosanoid

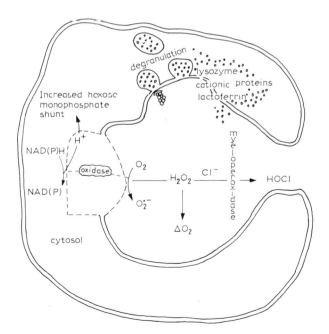

Fig. 2.4. The activated neutrophil.

metabolism, they are intermediates and/or products in enzyme-cata-
lysed reactions, they are part of the cascade of events in tissue response
to invading microorganisms, they are regulatory molecules in biochemical processes. However, free radicals can become highly destruc-
tive to cells and tissues if their production is not tightly controlled.

Some biologically important oxidants are not radicals, for example,
hydrogen peroxide, hypochlorous acid, ozone and singlet oxygen.
Nevertheless, they can produce oxidative damage when attacking biological material and thus are sometimes referred to, with oxygen-
centred radicals, as 'reactive-oxygen species'. The body has antioxidant defences to cope with an excess of free radicals generated. However, when there is an imbalance between the free radicals generated
and the protective mechanisms that remove them, then excessive radical production can be damaging.

2.4.1. Phagocytosis

Neutrophils (the major component of phagocytic cells in the human
bloodstream) and macrophages are activated and undergo a respiratory burst when they encounter foreign particles (Fig. 2.4). The cell
flows around the foreign particle and engulfs it in a plasma membrane
vesicle. Oxygen uptake in neutrophils and macrophages is due to the
action of an NADPH-oxidase complex associated with the plasma
membrane, the electrons released on oxidation of NADPH reducing
oxygen to superoxide radical. The engulfed foreign particles in the
plasma membrane vesicle, therefore, are exposed to a high flux of
superoxide radicals in the phagocyte cytoplasm, some of which are
also released extracellularly. The reaction products of superoxide ions
are believed to be partly responsible for the removal and destruction
of bacteria and damaged cells (Babior, 1978).

In view of its low reactivity it is unlikely that superoxide itself is
responsible for killing the invading material (although some strains of
bacteria are quickly killed by hydrogen peroxide formed by the dis-
mutation reaction [2.21]). Once the phagocytic vacuole is formed, fu-
sion with other granules in the neutrophil cytoplasm releases myelo-

peroxidase which utilizes hydrogen peroxide as a substrate and oxidizes chloride to hypochlorous acid [2.22]. The latter can oxidize many biological molecules, especially reduced thiol groups. This leads to a series of events resulting in bacterial killing.

$$2H^+ + 2O_2\cdot^- \xrightarrow{\text{superoxide dismutase}} H_2O_2 + O_2 \qquad [2.21]$$

$$H_2O_2 + Cl^- \xrightarrow{\text{myeloperoxidase}} HOCl + OH^- \qquad [2.22]$$

The relative importance of the contribution of superoxide/hydrogen peroxide and hypochlorous acid in the bacterial killing mechanism is seen in patients with chronic granulomatous disease (CGD, with a defective NADPH–oxidase system), and those that are myeloperoxidase-deficient. CGD patients show persistent multiple infections especially in the skin, lungs, liver and bones by those bacterial strains whose killing by neutrophils requires oxygen. Individuals who are deficient in myeloperoxidase show no symptoms.

2.4.2. Eicosanoid metabolism

Arachidonic acid metabolism through eicosanoid biosynthesis is accompanied by the generation and subsequent utilization of oxygen-derived free radicals. Arachidonic acid, released from lipids as a result

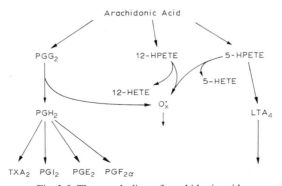

Fig. 2.5. The metabolism of arachidonic acid.

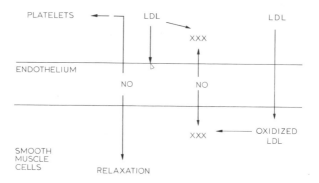

Fig. 2.6. Postulated modes of action of nitric oxide. Schematic diagram of the effects of nitric oxide on platelets and smooth muscle cells and the hypothesized interactions with low-density lipoproteins (LDL).

of activation of phospholipases by tissue injury or by hormones, may be metabolized by the prostaglandin or leukotriene pathways (Fig. 2.5). The peroxidase-catalysed conversion of PGG_2 to PGH_2 (Lands, 1979) and the mechanism of hydroperoxy fatty acid to the hydroxy fatty acid conversion both yield oxygen radicals (Miyamoto et al., 1976; Kuehl et al., 1977; Van der Ouderra et al., 1977; Yamamoto et al., 1980). These radicals can be detected by ESR spectroscopy at low temperatures and can be captured by antioxidant compounds with consequent stimulation of PGG_2 to PGH_2 and subsequent prostanoids. One of the consequences of scavenging the reactive oxygen species is to prevent the irreversible degradation of the cycloxygenase enzyme (Hemler and Lands, 1980; Yamamoto et al., 1980). The radical species, depicted as $[O_x^-]$ has many of the properties of the hydroxyl radical, although it is known not to be this species. It has been postulated that an active iron-oxygen species such as a ferryl derivative is formed in the peroxidase reaction (Kuehl et al., 1977). It has also been proposed that the haem iron in the cyclooxygenase enzyme must be reduced for maximal activity (Peterson et al., 1981) and that superoxide radical, lipid hydroperoxides and hydrogen peroxide can satisfy this criterion. Hence the stimulatory effects of hydroperoxides and

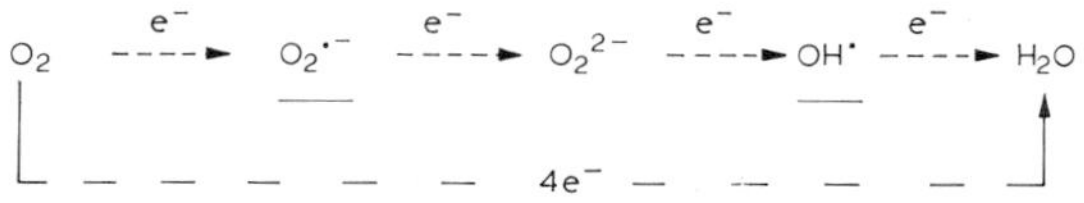

$$O_2 \; ----\blacktriangleright \; O_2^{\cdot -} \; \overset{e^-}{----\blacktriangleright} \; O_2^{2-} \; \overset{e^-}{----\blacktriangleright} \; OH^{\cdot} \; \overset{e^-}{----\blacktriangleright} \; H_2O$$

Fig. 2.7. The sequential reduction of oxygen via superoxide radical, peroxide and hydroxyl radical to water.

hydrogen peroxide on the activity of this enzyme can be understood. The inhibition of these reactions may have effects on the overall functions of arachidonic acid metabolism. Inhibition of prostaglandin production at the cycloxygenase part of the pathway may lead to overproduction of leukotrienes, hydro(per)oxy fatty acids, e.g., 5- or 12-HETE and peroxidase-derived free radicals all of which will have profound pathophysiological effects (Deby and Deby-Dupont, 1980).

2.4.3. Endothelial-derived relaxing factor NO

For some time a species called endothelial-derived relaxing factor, defied identification. However, it is now generally agreed that this factor is the simple radical, NO (see Chapter 1, section 1.3). This acts by inducing the relaxation of smooth muscle cells, through a process involving the activation of guanylate cyclase, and inhibits the aggregation of platelets (Bruckdorfer, 1989) (Fig. 2.6). In the absence of oxygen, it is a relatively long-lived species with a half-life of the order of 6 s in biological systems.

2.5. Toxicity of oxygen

The toxicity of oxygen is related to its high electron affinity, producing a variety of potentially damaging intermediates (Fig. 2.7): superoxide, hydrogen peroxide, hydroxyl radical, and to its tendency to add to radicals ($R^{\cdot}$) to give peroxyl radicals ($ROO^{\cdot}$). It can also be involved in singlet oxygen production.

A major agent and mediator of oxygen toxicity is the superoxide radical. Superoxide radical may be formed in vivo (Fridovich, 1983, 1989) in a number of ways:

1. The major source is the activity of the mitochondria and microsomal electron transport chains. The 4-electron reduction of oxygen to water, of course, is the normal process underlying mitochondrial electron transport:

$$O_2 + 4H^+ + 4e^- \rightarrow 2H_2O \qquad [2.23]$$

Cytochrome oxidase keeps all the partially reduced oxygen intermediates tightly bound to its active centre, and in general there is little leakage of electrons. However, some other components of the electron transport chain can leak electrons directly onto oxygen. Since oxygen can accept one electron at a time, superoxide radicals are released, production increasing with increasing oxygen concentration.

2. The respiratory burst of neutrophils when they come into contact with foreign particles (see Section 2.4.1).

3. Endothelial cells, macrophages, fibroblasts have all been reported to generate superoxide radicals.

Superoxide dismutase is present in all aerobic cells and reduces superoxide to hydrogen peroxide. Hydrogen peroxide is not a radical and is generally relatively unreactive. Its cytotoxicity is partly due to the damage done when hydrogen peroxide comes into contact with iron or copper in a catalytic form when hydroxyl radicals may be generated (see Chapter 1). Iron or copper bound to transport or functional proteins does not react with H_2O_2 or only very slowly, although haem-containing proteins may under certain conditions react quite readily.

Superoxide and hydrogen peroxide have been implicated in the regulation of cell proliferation (Burdon and Rice-Evans, 1989), perhaps by signalling between cells.

Singlet oxygen is used in the treatment of cancer by photosensitization (photodynamic therapy). This involves treatment with certain porphyrin derivatives which localize preferentially in the tumour, followed by irradiation with ultraviolet light. Singlet oxygen is thereby generated and is thought to damage the tumour vasculature which leads to tumour cell death.

2.5.1. Important determining factors for the way in which individuals respond to oxygen

The signs of oxygen toxicity depend on the organism under study, in terms of its age, physiological state and nutritional status. Factors involved include the presence of varying amounts of vitamins A, E and C, the trace elements manganese, copper, zinc and selenium. Limited intake of iron, other antioxidants (added to foodstuffs), polyunsaturated fatty acid intake may all be important. For example, newborn rats survive in pure oxygen much longer than adult rats which quickly develop respiratory stress; in contrast, adult humans can breathe pure O_2 for several hours without much damage, although prolonged exposure produces alveolar damage and eventual lung fibrosis. High-pressure oxygen causes acute toxicity to the central nervous system in humans, leading to convulsions – this has been a problem in diving and must be taken into account using hyperbaric oxygen therapy, in the treatment of cancer and lung diseases.

2.6. Pathological systems and toxic responses

There is considerable debate concerning the role played by free radical reactions, protein oxidation, DNA damage and lipid peroxidation in human disease and toxicology. Radical species have indeed been implicated in many disease states (Table 2.1). The question as to whether free radicals are a major cause of tissue damage in human disease or an accompaniment to or a consequence of such injury is by no means clear in many instances. However, what is clear is that diseased or damaged tissues undergo radical reactions more readily than normal, which may exacerbate the primary lesion.

The reasons for the increased oxidizability of damaged tissues in-

TABLE 2.1

Some of the disease states in which free radicals have been implicated

Inflammatory disorders	Reperfusion injury
Rheumatoid arthritis	Atherosclerosis
Alcoholism	Lung disorders
Iron overload	Tumour promotion
Certain blood disorders	Parkinson's disease
Retinopathy of prematurity	Porphyria
Toxic liver injury	

(Paracetamol, adriamycin, quinones, nitrosamines, aromatic amines, halothane etc. form free-radical intermediates by metabolic activation in the liver.)

clude inactivation of some antioxidants, the leakage of antioxidants from cells, release of metal ions from storage sites and from metalloproteins hydrolysed by enzymes released from, for example, damaged lysosomes. In myocardial reperfusion injury, the reintroduction of oxygen to the ischaemic tissue generates radicals which worsen the injury; in atherosclerosis, enhanced peroxidation in the atherosclerotic lesion leads to further foam cell formation and growth of the lesion (Steinberg et al., 1989); in traumatic injury to the brain and spinal cord, there is evidence suggesting that iron released into the surrounding area stimulates further radical-mediated damage (Halliwell, 1990). Thus each set of evidence proposing the involvement of free radicals as important contributors to the pathology of a given disease must be judged on the accuracy and specificity of the methodology and techniques utilized for measuring these processes in organs, tissues and cells.

2.6.1. Factors influencing the release of free radicals in cells and tissues

Free radicals may arise as a consequence of increased radical generation in a variety of ways or from decreased efficacy of the natural protective mechanisms, thus limiting the body's normal control of the endogenously generated radical species (Fig. 2.8).

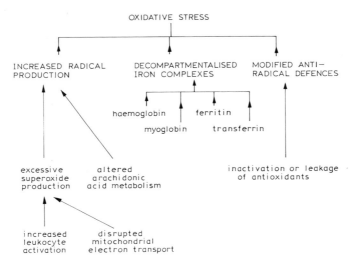

Fig. 2.8. Factors controlling the production of free radicals in cells and tissues (Rice-Evans, 1990a). Free radicals may be generated in cells and tissues through increased radical input mediated by the disruption of internal processes or by external influences, or as a consequence of decreased protective capacity. Increased radical input may arise through excessive leukocyte activation, disrupted mitochondrial electron transport or altered arachidonic acid metabolism. Delocalization or redistribution of transition metal ion complexes may also induce oxidative stress, for example, microbleeding in the brain, in the eye, in the rheumatoid joint. In addition, reduced activities or levels of protectant enzymes, destruction or suppressed production of nucleotide coenzymes, reduced levels of antioxidants, abnormal glutathione metabolism, or leakage of antioxidants through damaged membranes, can all contribute to oxidative stress.

2.6.2. How do free radicals damage cells and tissues?

Among the major cellular and extracellular targets for reactive radical species are proteins, unsaturated fatty acyl components of lipids and lipoproteins, and DNA constituents including carbohydrates, as is depicted in Fig. 2.9.

2.6.2.1. Vulnerability of lipids to oxidative damage
Recent review articles (Halliwell and Gutteridge, 1984; Slater, 1984; Kappus, 1985) have concluded that oxygen-derived free radical spe-

Fig. 2.9. Targets of free radical damage.

cies such as superoxide ($O_2^{\cdot-}$) and hydroxyl radicals (˙OH) as well as hydrogen peroxide are important mediators of cellular injury via the destruction of membranes or alteration of critical enzyme systems. The phospholipid component of cellular membranes is a highly vulnerable target due to the susceptibility of its polyunsaturated fatty acid side-chains to peroxidation. This can lead to changes in the membrane permeability characteristics and the ability to maintain transmembrane ionic gradients (Slater, 1984). The peroxidation of polyunsaturated fatty acid side-chains of the membrane lipids is a consequence of many types of cellular injury in which free radicals intermediates are produced in excess of local defence mechanisms and has been extensively reviewed (Halliwell and Gutteridge, 1985).

Lipid peroxidation may be initiated by any primary free radical which has sufficient reactivity to extract a hydrogen atom (Fig. 2.10) from a reactive methylene group of an unsaturated fatty acid. For example, species such as hydroxyl radicals ˙OH, alkoxyl radicals RO˙, peroxyl radicals ROO˙ and alkyl radicals R˙ may be involved. The formation of the initiating species is accompanied by bond rearrangement that results in stabilization by diene conjugate formation. The lipid radical then takes up oxygen to form the peroxyl radical. Peroxyl radicals can combine with each other or they can attack membrane proteins, but they are also capable of abstracting hydrogen from adjacent fatty acid side-chains in a membrane and so propagate the chain reaction of lipid peroxidation. Hence a single initiation event can result in the conversion of hundreds of fatty acid side-chains into lipid monohydroperoxides. This propagation phase can be repeated many times. Thus, an initial event triggering lipid peroxidation can be amplified, as long as oxygen supplies and unoxidized fatty acid chains are available. The length of the propagation chain also depends on the lipid/protein ratio in the membrane (the chance of a radical interacting with a membrane protein) will clearly increase as the protein content of the membrane rises. The presence in the membrane of chain-breaking antioxidants will also interrupt the propagation phase. Thus interception by a chain-breaking antioxidant might lead to a radical species

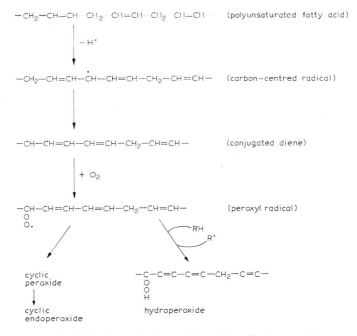

Fig. 2.10. Lipid peroxidation and formation of hydroperoxides.

capable of reacting with another peroxyl radical, it might disappear harmlessy (for example, by dimerisation) or it might be regenerated as a functioning antioxidant by reaction with another molecule. For example, it has been proposed that the α-tocopherol radical can be converted back to tocopherol by a reaction with ascorbate at the surface of the biological membrane (McKay, 1985; Burton and Ingold, 1986).

Lipid hydroperoxides are fairly stable molecules under physiological conditions, but their decomposition is catalysed by transition metals and metal complexes (O'Brien, 1969). Both iron(II) and iron(III) are effective catalysts for hydroperoxide degradation, but the former is more so (Halliwell and Gutteridge, 1984). These include complexes of iron salts with low molecular weight chelates, non-haem iron proteins, free haem, haemoglobin, myoglobin.

Reduced metal complexes react with lipid hydroperoxides (LOOH) to give alkoxyl radicals (LO$^\bullet$):

$$LOOH + Fe^{2+}\text{-complex} \rightarrow LO^\bullet + Fe^{3+}\text{-complex} + OH^- \qquad [2.24]$$

Oxidized iron complexes react more slowly to produce alkoxyl and peroxyl radicals (Davies and Slater, 1987) and, under certain conditions, ferryl complexes (Labeque and Marnett, 1988):

$$LOOH + Fe^{3+}\text{-complex} \longrightarrow LO^\bullet + [Fe\,(IV){=}O]^{2+}\text{-complex} + H^+$$
$$\searrow LOO^\bullet + Fe^{2+}\text{-complex} + H^+ \qquad [2.25]$$

Alkoxyl and peroxyl radicals can initiate new rounds of lipid peroxidation and propagate further radical chain reactions thus amplifying the initial damage. Cleavage of the carbon bonds during lipid peroxidation reactions (Fig. 2.11) results in the formation of:

1. Alkanals such as malonaldehyde (Tappel, 1980) which interact with protein thiols, cross-link amino acid groups of lipids and proteins, and give rise to chromolipids and aggregated proteins (Fig. 2.12).

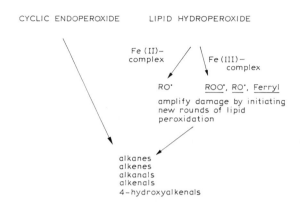

Fig. 2.11. Products of cleavage of carbon bonds during lipid peroxidation.

2. Alkenals, such as 4-hydroxynonenal (Esterbauer, 1985) which is very biologically active, inhibiting platelet aggregation, activating adenyl cyclase activity (Esterbauer, 1985) and is a substrate for glutathione transferases; and

3. Alkanes, e.g., pentane, produced by this mechanism as an end-product of the oxidation of linoleic and arachidonic acid, and ethane from linolenic acid (Fig. 2.13) (Tappel and Dillard, 1981).

2.6.2.2. Delocalization of lipid damage

The production of highly reactive free radicals leads to primary reactions close to the site of formation. It will not diffuse far before it interacts within its micro-environment which in a membrane would include the initiation of lipid peroxidation. Secondary products of such

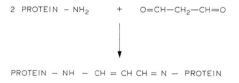

Fig. 2.12. Interaction between malonaldehyde and protein amino groups.

peroxidative events such as lipid peroxyl radicals and lipid hydroperoxides may diffuse in the plane of the membrane before reacting further, thereby spreading the biochemical lesion. Such processes therefore not only affect the structural and functional integrity of the
membrane, its fluidity and permeability, but also the breakdown
products of lipid peroxidation can further damage cellular function.
For example, alkenals, alkanals, lipid hydroperoxides may undergo
degradative reactions and be metabolized rapidly; some of these, e.g.,

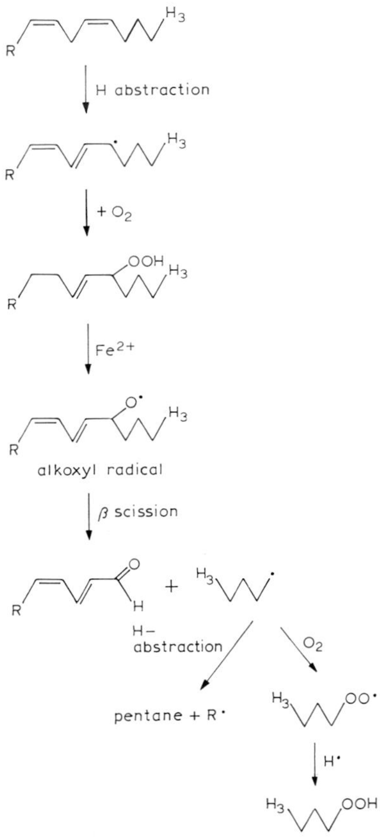

Fig. 2.13. Formation of alkanes as metabolites of lipid hydroperoxides.

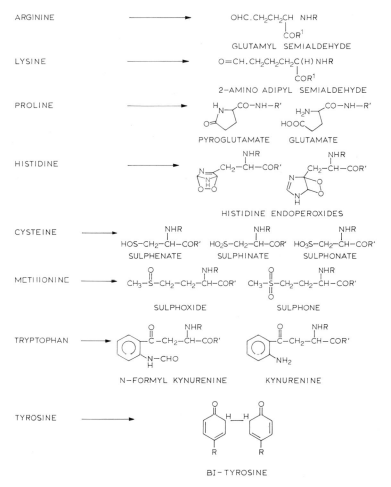

Scheme 2.2. Modifications to amino acid side-chains in proteins after free radical attack.

the lower molecular weight hydroperoxides, aldehydes and 4-hydroxyalkenals, can escape from the membrane and cause disturbances at a distance. Therefore a reaction that originally produces a radical which interacts within its microenvironment may produce a sequence of later events that direct disturbances throughout the cell, its membrane and, in some instances, into the extracellular domain.

2.6.2.3. *Vulnerability of proteins to radical-mediated damage*

Proteins are critical targets for free radical attack, both intracellularly and extra-cellularly and, because many are catalytic, modifications may have an amplified effect.

Several amino acyl constituents crucial for the protein's function are particularly vulnerable to radical damage (Roshchupkin et al., 1979; Singh et al., 1982; Sies, 1986), as shown in Scheme 2.2 and Table 2.2, with the consequences of oxidative modification. It is generally accepted that reactive oxygen species can react directly at several of these sites on a protein; in addition, in some instances, when protein radicals are formed at a specific amino acyl site, they can be rapidly transferred to other sites within the protein infrastructure, the pathway so far proven being illustrated in Fig. 2.14 (Butler et al., 1988).

TABLE 2.2

Products formed after modifications of amino acids induced by free radicals (Rice-Evans, 1990b).

Arginine	• Glutamic semialdehyde + NO
Lysine	• 2-Amino adipylsemialdehyde
Proline	• Glutamic semialdehyde →pyroglutamate→glutamate
Histidine	• Histidine endoperoxides
	• Aspartate, asparagine
Cysteine	• Cys-disulphides mixed-disulphides

	• Sulphenic acid, sulphinic acid,
	  Sulphonic acid
	  (via alkyl thioradicals)

Methionine	• Methionine sulphoxide Methionine sulphone
Tryptophan	• 5-Hydroxy tryptophan
	• Kynurenine
	• *N*-Formylkynurenine
Tyrosine	• Bityrosine (not in the presence of O_2, $O_2^{\cdot-}$)
Phenylalanine	• Tyrosine (in presence of $^\cdot$OH)

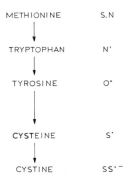

Fig. 2.14. Radical transfer within proteins with associated radical centres. Once free radicals are produced in a protein they can be rapidly transferred to other sites within the protein infrastructure through this route, so far proven. (Adapted from Butler et al., 1988.)

Proteins are also particularly susceptible to attack from radical intermediates of lipid peroxidation, alkoxyl LO˙ and peroxyl LOO˙ radicals. These may react with proteins closely associated with the peroxidizing lipids. Defined radical species may have specific effects on particular amino acid side-chains; for example, methionine oxidation to methionine sulphoxide and cysteine to cysteic acid may be mediated by superoxide radicals; oxidation of tryptophan to kynurenine, N-formylkynurenine, 5-hydroxytryptophan (Butler et al., 1988) may reflect direct attack by hydroxyl radicals or by peroxyl radicals formed as metabolites of adjacent lipid hydroperoxides in the membrane; lysine may be modified by stable products of lipid peroxidation such as malonyldialdehyde or 4-hydroxynonenal. The consequences of such damage may be aggregation and cross-linking or protein degradation and fragmentation depending on the nature of the vulnerable protein component and on the attacking radical species (Wolff et al., 1986; Wolff and Dean, 1986). Radical-mediated protein breakdown has been shown to be an earlier event than lipid peroxidation (Davies and Goldberg, 1987).

The consequences of oxidative modification to proteins may be altered enzymic activity and altered membrane and cellular function resulting from degradation or cross-linking. Such damage to the membrane transport proteins, for example, might affect the ionic homeostasis of the cells leading to calcium accumulation. The resulting potential for activation of phospholipases, proteases, accumulation of mitochondrial calcium may lead to extensive membrane and cellular deterioration and gross exacerbation of the initial lesion.

Oxidative modification of proteins renders them more susceptible to proteolytic attack and enzymic hydrolysis (Wolff et al., 1986; Davies, 1987). Hence, if generation of active oxygen species occurs significantly in vivo, one consequence may be accelerated hydrolysis of damaged proteins. Therefore, radical generation at inappropriate sites may lead to destruction of the protein and pathological tissue degradation.

In addition to cellular reductants and antioxidants whose functions are to replenish the system after radical attack and cleave protein disulphides, repair mechanisms may exist in certain tissues for dam-

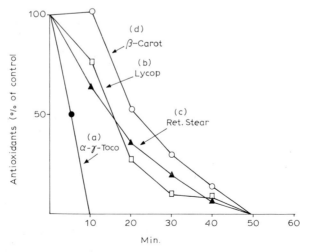

Fig. 2.15. Relative importance of plasma antioxidants. Time-course of the consumption of antioxidants during oxidation of plasma low-density lipoproteins: (a) tocopherols, (b) lycopene, (c) retinyl stearate, (d) β-carotene. (Adapted from Esterbauer et al., 1989.)

TABLE 2.3
Antioxidant defences

1. Extracellular antioxidant defences in human plasma

TRANSFERRIN (1.2–2.0 mg/ml)	Apotransferrin binds iron(III) for transport and de-livery to cells. It is in its capacity as an iron-binding protein which renders it also able to function as an antioxidant by making iron(III) unavailable for par-ticipation in iron-catalysed radical reactions. Only about 30% of the iron-binding sites on the transferrin in human plasma are normally occupied in vivo.
CAERULOPLASMIN (0.2–0.4 mg/ml)	This copper-containing protein is regarded as a phys-iological inhibitor of lipid peroxidation. In this one of its many roles, it acts as an antioxidant by virtue of its ferroxidase activity, converting iron(II) to iron-(III) by electron transfer.
ALBUMIN (50–60 mg/ml)	Albumin, one of the most important proteins in human plasma, is able to bind copper(II) tightly and iron weakly. Copper(II) bound to albumin is still ef-fective in generating radical species in the presence of hydrogen peroxide. Thus macromolecules function-ing by this mechanism are called sacrificial antioxi-dants, since the hydroxyl radical is generated locally on the protein and reacts at the specific site (Halli-well, 1988). The binding of copper ions to albumin may represent a protective mechanism overall, since the damaged albumin can be quickly replaced.
HAPTOGLOBIN/HAEMOPEXIN	These proteins bind free haemoglobin/haem, thus protecting delocalised haemoglobin from influences which might otherwise activate the protein or desta-bilize the haem ring and promote iron release.
URATE	Urate has the potential for chelating iron and copper rendering them unreactive and thus inhibiting lipid peroxidation. It also reacts with singlet oxygen (Ames et al., 1981).
α-TOCOPHEROL (10–20 μg/ml)	This is a membrane-bound (lipid-soluble) chain-breaking antioxidant which reacts with peroxy and other reactive radicals (Diplock, 1983; Burton and Ingold, 1986; Esterbauer et al., 1988); it is the major lipid-soluble chain-breaking antioxidant in human plasma, bound to lipoproteins (approximately 7/LDL molecule (Esterbauer et al., 1989)).

TABLE 2.3 (*continued*)

ASCORBATE (0.006–0.017 mM)	Although at relatively low levels in human plasma, many tissues (lens, lung, brain, heart) contain ascorbate in millimolar concentrations. This antioxidant nutrient scavenges singlet oxygen and reacts rapidly with $\cdot OH$, $O_2^{\cdot -}$ and $HO_2^{\cdot}$. It has also been proposed that it acts synergistically with vitamin E. Reaction of ascorbate with superoxide radical, for example, produces the semidehydroascorbate radical which is relatively stable (Bielski et al., 1975). Ascorbate is reported to be the most effective aqueous-phase antioxidant in human blood plasma (Fig. 2.15) (Frei et al., 1989) protecting against oxidants released from polymorphonuclear leukocytes and against lipid-soluble peroxyl radicals.
β-CAROTENE (v.low concentrations)	In the core of lipoproteins there is approximately 1 carotene/3LDL (Esterbauer et al., 1989). This carotenoid scavenges singlet oxygen and peroxyl radicals (Burton and Ingold, 1984; Vile and Winterbourn, 1988).
LYCOPENE	Lycopene is a low concentration, core carotenoid antioxidant in lipoproteins, approximately 1 lycopene/5LDL (Esterbauer et al., 1989; Di Mascio et al., 1989).
METALLOTHIONEIN	The metallothioneins are generated in response to high concentrations of certain metal ions. They bind these ions extremely efficiently such that they are rendered harmless and can be removed from the body.
BILIRUBIN	Bilirubin has been proposed to be an effective antioxidant in terms of its capacity to protect albumin-bound polyunsaturated fatty acids (Stocker et al., 1987).

2. Intracellular antioxidant defences

SUPEROXIDE DISMUTASE	The major intracellular antioxidant enzyme in aerobic cells, the Cu-Zn form in the cytoplasm and the manganese form in the mitochondria, reduces superoxide radicals to hydrogen peroxide very rapidly and specifically.

TABLE 2.3 (*continued*)

GLUTATHIONE PEROXIDASE/ GLUTATHIONE TRANSFERASES	Selenium-containing glutathione peroxidase is essential for removing hydrogen peroxide, reducing it to water at the expense of reducing equivalents donated by GSH (Cohen and Hochstein, 1963) and for removing lipid hydroperoxides after cleavage from membranes.

$$2\,GSH + H_2O_2 \;\to\; GS\text{-}SG + 2H_2O$$
$$LOOH \to GS\text{-}SG + LOH + H_2O$$

Selenium is normally found at levels of 120–200 ng/ml in whole blood.

Selenium deficiency (i.e., glutathione peroxidase deficiency) in animals produces a variety of diseases that are strikingly similar to those induced by vitamin E deficiency.

Selenium-independent glutathione peroxidase or glutathione transferase also utilizes reduced glutathione (see Section 6.3).

Thus, maintenance of the reduced glutathione levels is essential for supporting the activity of this enzyme (as, indeed, for other enzymes). Cells are thus equipped with NADPH-dependent glutathione reductase to reduce the oxidized glutathione.

$$GS\text{-}SG \xrightarrow[+\ NADPH]{\text{glutathione reductase}} 2GSH$$

CATALASE

Catalase reacts very rapidly with hydrogen peroxide, converting it to water. It is located in the cytoplasm of erythrocytes but compartmentalized in the peroxisomes of most other cells.

Catalase is especially concentrated in liver and erythrocytes, but is low in brain, heart and skeletal muscle. The normal low concentrations of hydrogen peroxide production are efficiently reduced in cells by glutathione peroxidase. However, if the concentration of hydrogen peroxide is raised, catalase becomes important.

3. Secondary protection

In addition, cells contain mechanisms for repairing DNA after an attack by radicals. Systems for degrading proteins damaged by free radicals (Marcillat et al., 1988) and erythrocytes contain oxyhaemoglobin which is capable of protecting the cell membranes against peroxidative damage (Rice-Evans et al., 1985).

aged protein components (Sies, 1986) such as methionine sulphoxide reductase in the human lens, the lung and the neutrophil.

2.7. Antioxidant defences

In the normal course of events cells have adequate antiradical defence mechanisms, both those synthesized in vivo and those taken up in the diet. The antioxidants synthesized in the body include a range of proteins, enzymes and transition metal-binding proteins, as well as those that are generated during the metabolism of other molecules (bilirubin and urate). It is only the levels of the antioxidant vitamins, ascorbate, tocopherol, carotene, lycopene, that can be controlled by dietary means. The antioxidants, which are located intracellularly, extracellularly and bound to the membrane are shown in Table 2.3 and Fig. 2.15. In general all the intracellular antioxidants are systems for removing reactive oxygen species, whereas among the extracellular components are those which additionally serve the essential function of keeping the transition metal catalysts under control. Any situation which increases the turnover of the antioxidant cycle, whether increased oxidative stress or modified antiradical defences, can lead to progressive cellular and membrane damage.

CHAPTER 3

The detection and characterization
of free radical species

3.1. Electron spin resonance (ESR) or Electron
paramagnetic resonance (EPR) spectroscopy

This technique, alternatively described as electron paramagnetic resonance (EPR), is unique in that it is only sensitive to transitions involving unpaired electrons. There are a variety of reviews and books on the subject (Alger, 1968; Atherton, 1973; Symons, 1978) including some new texts. In particular, the widely used book by Wertz and Bolton (1972) is being extensively updated.

Here we are concerned only with painting a broad impressionistic picture of the technique itself, together with some general comments on its method and ease of use. The technique, developed just after the second world war, using the predictions of van Vleck, was originally used to study paramagnetic transition metal complexes (Abragam and Bleaney, 1970). It has more recently been used to study radicals, although it remains very important in the field of metal complexes.

It is closely related to the analytically more important technique of nuclear magnetic resonance (NMR) and we draw comparisons widely herein, since NMR will be much more widely understood by our readers.

3.1.1. The spectroscopic act

Most forms of spectroscopy depend on coupling between the electric vector of electromagnetic radiation and an electrical charge in the molecule, such as the movement of an electron or a change in the di-

51

pole moment. In ESR and NMR spectroscopy, it is the magnetic component that is involved. Separation between the magnetic states is achieved by the application of a magnetic field (B) as indicated in Fig. 3.1. The sample is placed in the region of high magnetic field in a microwave cavity sustaining stationary microwaves of relatively low power, and the DC magnetic field is increased (with a superimposed modulation) until resonance is achieved (Fig. 3.1). The coupling can be thought of as a spin-flip, the electron spin moving from the $+1/2$ to the $-1/2$ state in the magnetic field. (Spins are flipped both ways, a net absorption of power occurring because the lower state is more populated.)

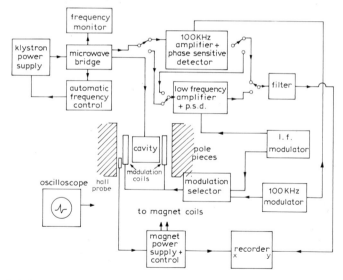

Fig. 3.1. Schematic diagram for a standard C.W. X-band spectrometer. The sample is placed vertically into the centre of the cavity in the region of maximum microwave magnetic field. Various forms of stationary waves are set up in the cavity, depending on its shape. The basis of the instrument is the Klystron as a source of monochromatic microwaves, the permanent electromagnet, giving a homogeneous field through the microwave cavity (or resonator). The field is swept to generate the spectrum, and the modulation coils provide a rapid sampling (often at ca. 100 kHz) to give phase-sensitive detection.

3.1.2. The ESR spectrometer

The basic instrument comprises a microwave source, a cavity resonator (or loop-gap resonator), an amplifier and a recording system. A
simple scheme is given in Fig. 3.2. Almost all spectrometers operate
at X-band frequencies (ca. 9000 MHz), using a monochromatic Klystron source which is only tunable over a small range. Hence, to obtain
resonance, the static magnetic field is swept. Generally there are mod-
ulating coils which operate at high frequencies (commonly ca. 100
MHz). Obviously, modern spectrometers incorporate a multitude of
sophisticated addenda, aimed at increasing the sensitivity, maintaining the tuning over long periods and increasing the ease of operation.
Dedicated computers are the norm, so that spectra can be accumulat-

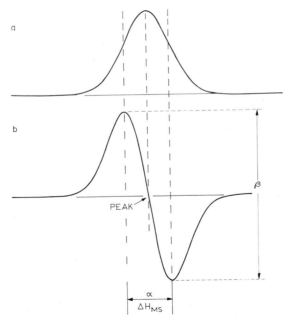

Fig. 3.2. (a) Absorption band whose first derivative is shown in (b), which is the normal
mode of plotting an ESR spectrum. Note that the width between points of maximum
slope ($\Delta H_{MS}\alpha$) is easily measured and is the standard representation of band width.

ed to average out noise and stored for various forms of adjustment. (Base-line changes, accurate g- and A-value recording, spectral additions and subtractions, resolution enhancement etc.) There is little point in elaboration here, since extensive details are given in instruction manuals and in advanced ESR texts (Poole, 1967).

A particular problem arises when aqueous solutions are studied because of the high dielectic absorption of water in the microwave region. In order to prevent extensive microwave absorption, and hence loss of sensitivity, a region of the cavity is selected where the electric component is low but the magnetic component is high. For a standard H_{o12} cavity, thin cells holding ca. 0.1 ml are used. These can be made demountable for tissue studies or can be part of a flow system.

As with NMR spectroscopy it is sometimes helpful to measure spectra over a range of frequencies. The most popular second choice is Q-band (ca. 35 000 MHz, i.e., ca. 3.5-times X-band frequencies). In principle, going to higher frequencies should increase the sensitivity, because of the Boltzmann factor, but in practice this is about compensated by the small size of the cavity resonator, and hence the need for small samples. This can, of course, be a great advantage if only small samples are available, as, for example, is often the case if isolated metallo enzymes are being studied. The next most popular frequency is S-band (ca. 3000 MHz, or ca. 1/3 that of the X-band). Sensitivities are lower, but for aqueous systems, using loop-gap resonators, this may not be a serious factor.

Although sensitivity decreases as the frequency is reduced, there is one major advantage of going right down to radio frequencies, namely that large volumes of aqueous solutions can now be used, in contrast with the situation at X-band. For example, such a spectrometer (300 MHz) constructed at Leicester (Brivati et al., 1990) gives a good response using a 250 ml aqueous sample, and this volume can be readily increased. Using optimized conditions, the same solution of a nitroxide radical has a signal-to-noise ratio only ca. 1/6 less favourable than that at X-band. These spectrometers can be used for whole-body ESR imaging which, although in its infancy, promises much.

There are two more sophisticated ESR techniques, which are of major importance in biological studies, namely ENDOR (electron-nuclear double resonance) spectroscopy and various forms of spin-echo ESR spectroscopy sometimes coupled with Fourier-transform tchniques. The former follows changes in the ESR spectrum during irradiation with power at the hyperfine frequencies (in addition to the microwave power). When high microwave powers are used so that the ESR spectrum is partially saturated, nuclear relaxation induced by the swept radiofrequency radiation causes a change in the ESR response as the hyperfine frequency is traversed. The sensitivity is usually considerably reduced compared with normal ESR spectroscopy, but lines are often very narrow, so that hyperfine features are resolved which are lost in the broad ESR lines. This technique is mainly used to help unravel very complex liquid-phase ESR spectra to pick up unresolved hyperfine features (of particular use with spin-traps) and to give greater details of small coupling constants for transition metal complexes (such as metallo-proteins) in the solid-state.

Spin-echo methods are less widely used, but certainly rival the ENDOR method, and the sensitivity can be much greater. The method has been most widely developed in studies of transient species produced in flash photolysis or pulse radiolysis experiments. At this stage, the great advantages found in NMR spectroscopy have not been fully realized in ESR spectroscopy, but time will tell.

3.1.3. Spectroscopic parameters

The main parameters derived from ESR spectra are the g-values (analagous to the chemical shift in NMR), the hyperfine coupling constants (like spin-spin coupling constants), line-widths and intensities. Because spectra are usually displayed as first derivatives (Fig. 3.2), line-widths are conveniently measured as the distances between points of maximum and minimum slope (α in Fig. 3.2). Intensities are often quoted as the total extent of each line (β in Fig. 3.2), but this can be misleading, and it sometimes helps to integrate to give the normal absorption spectrum.

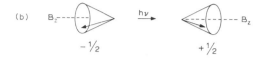

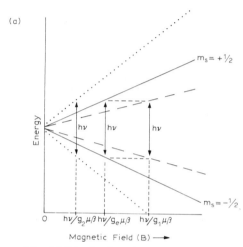

Fig. 3.3. Divergence of the $\pm 1/2$ levels of an electron in a radical as the external magnetic field (B) is increased. If the full lines represent the divergence for normal radicals with $g = 2$, the _____ lines are for a species with $g < 2$ and the ..... lines for a species with $g > 2$; (b) shows how the precessing electron flips its orientation on absorbing $h\nu$ of energy.

g-values. The g-value gives the rate of divergence of the levels, as shown in Fig. 3.3. Rapid divergence gives a high g-value and a low-field resonance, and vice versa. When only the electron spin contributes to the total magnetism, $g = 2.0023$, the 'free-spin' value. This is given by:

$$h\nu = g\mu_\beta B$$

where ν is the microwave frequency (fixed); B is the applied magnetic field (varied) and μ_β, the Böhr magneton, is a constant. When the field induces orbital motion, the resulting magnetic moment adds to, or subtracts from that of the electron, giving shifts in the g-value from 2.0023. If coupling is via an empty orbital they subtract, giving low g-values, and if coupling is via a filled orbital they add, giving high g-values (Fig. 3.4).

3.1.4. Hyperfine coupling

For those familiar with NMR spectroscopy it may be helpful to realize that the ESR g-shift is comparable with the NMR chemical shift. Similarly, electron-nuclear hyperfine coupling can be compared with nuclear-nuclear spin-spin coupling in NMR. (In systems containing more than one unpaired electron per molecule, electron spin-electron spin coupling is, of course, important. For doublet-state radicals, this coupling does not arise: it is of great importance in triplet state molecules and in many high-spin transition metal complexes.)

To understand the origin of hyperfine coupling it must be recalled that many nuclei have net magnetic moments (which in diamagnetic compounds give rise to NMR spectra). Those with nuclear spin (I) of 1/2 taken up to ($\pm 1/2$) orientations in the resultant magnetic field, those with $I = 1$ have three choices ($+1, 0, -1$) etc. Consider a radical containing a single nucleus with $I = 1/2$. Half these radicals will have nuclei with $m_I = +1/2$ and the other half will have $m_I = -1/2$. Hence two resolved features are expected in the ESR spectrum, the

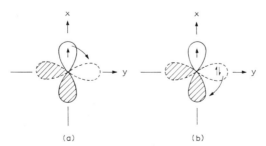

Fig. 3.4. Movement of an unpaired electron from one p-orbital (p_x) into another (p_y) induced by a magnetic field gives rise to orbital motion and hence a shift in the g-value: **(a)** coupling to a vacant orbital and **(b)** coupling to a filled orbital.

58 TECHNIQUES IN FREE RADICAL RESEARCH

separation (in first order) giving the hyperfine coupling (A) (Fig. 3.5). If $I = 1$, there are three possible orientations ($+1, 0, -1$) giving three ESR transitions, the separation again giving the hyperfine coupling.

Several coupled nuclei will make the pattern more complex. Thus one $I = 1/2$ nucleus and one $I = 1$ nucleus (e.g., ^1H and ^{14}N) will give 6 components (M_I values of $(1+1/2)$, $(1-1/2)$, $(0+1/2)$, $(0-1/2)$, $(-1+1/2)$, $(-1-1/2)$ from which both coupling constants can be derived. Since the magnetic moment of the electron is orders of magnitude greater than those of nuclei these splittings are generally small compared with those caused by the electron.

The link between ^1H coupling in NMR and ESR can be seen by comparing the ESR spectrum for $\cdot C(CH_3)_3$ radicals and $H\text{-}C(CH_3)_3$ molecules (looking at the unique proton only) – for the former, the methyl groups rotate rapidly, and the electron 'sees' 9 equivalent protons, giving a 10-line ESR spectrum of binomial relative intensities. Similarly, the unique proton couples to the 9 methyl protons, giving a similar 10-line NMR spectrum.

The task of interpreting complex isotropic ESR spectra resembles

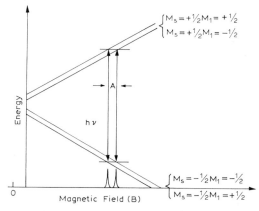

Fig. 3.5. Effect of hyperfine coupling to a single nucleus with $I = 1/2$ in the high-field region. At low magnetic fields, divergence is no longer linear and due allowance for this must be made. The ESR active transitions occur when only the electron-spin flips, so that only two transitions are indicated, separated by the hyperfine splitting.

that of NMR spectra, except that the complexity is generally less. It is a problem in pattern recognition and some people have no difficulties, whilst others may find the following suggestions helpful. The most important thing is to be sure that you start with the first line in the spectrum – this can be hidden in noise if there are many equivalent protons, so turn up the gain and average out the noise with a computer. Given that the first line is correct, measure always from this to other lines in the spectrum, as indicated in Fig. 3.6. The distance to line 2 (a_1) is the smallest hyperfine coupling. The distance to line 3 is exactly $2a_1$. Since line 2 is twice the intensity of 1 or 3, this is a 1:2:1 triplet, characteristic of two equivalent protons ($+1$, 0, -1). Now, measure from line 1 to line 4 to get a_2. Again this is twice the intensity of line 1, and is therefore the first line of a second triplet. Note that the first, small, triplet is now repeated starting with line 4. These are not *new* features. They merely repeat lines and hence are, in a sense, redundant. If you make a mistake, you will get a new splitting that is exactly equal to $a_1 - a_2$, so check that this has not occurred. The next feature, line 5, is equal in intensity to line 1, and the new splitting, a_3, does not repeat. Hence, this is the second component of a doublet.

We now have:

$a_1 = G(2H)$

$a_2 = G(2H)$

$a_3 = G(1H)$

This is a $3 \times 3 \times 2 = 18$ line spectrum and as can be seen (Fig. 3.6) accounts for 1/3 of the spectrum. So lines 1 – 18 are fully accounted for. Measuring from line 1 to line 19 gives a_4, and this repeats as a 1:1:1 pattern. Hence it must be due to ^{14}N nuclei ($I=1$). Each ^{14}N component comprises the same 18 line pattern, giving the total of 54 lines observed.

The proton splittings suggest an aromatic ring with a π electron coupling to two *ortho*-protons (a_2), two *meta*-protons (a_1) and a *para*-proton (a_4) nearly equal to the *ortho*-protons. These all relate to a ^{14}N substitution which is, in fact, a nitro group. Hence, as indicated in Fig. 3.6, the species must be the nitrobenzene negative-ion.

3.1.5. Anisotropies

So far we have only considered liquid-phase spectra, which give isotropic hyperfine splittings and the average g-values. However, ESR spectroscopists often study frozen samples (or, if available, single

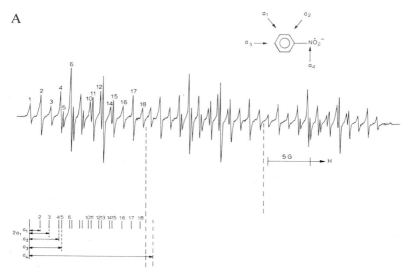

Fig. 3.6. **(A)** First derivative ESR spectrum assigned to nitrobenzene radical anions (dilute solution in dimethyl sulphoxide). To analyse the spectrum, measure always from line 1, as indicated. Line $1\rightarrow2$ gives a_1, and since $1\rightarrow3$ gives $2a_1$, lines 1,2 and 3 (1:2:1) clearly stem from two equivalent protons ($2H_1$). The splitting for $1\rightarrow4$ gives a_2 and this is clearly a similar 1:2:1 triplet, the third component being line 10 ($2H_2$). Line $1\rightarrow5$ gives a_3, assigned to a single proton (H_3). You have now accounted for the first 18 lines of the spectrum, because these splittings repeat themselves. To be sure that this is the case, either measure from line 1 to all the others in the set and see that the distances are combinations of a_1, a_2 and a_3 – or reconstruct the spectrum using the data established so far. The next new splitting is from line $1\rightarrow19$, giving a_4, and this clearly repeats, so that there are three sets of the 18-line pattern. These are roughly 1:1:1 and can therefore be assigned to ^{14}N coupling. The final assignments are in the insert. **(B)** Set of first-derivative ESR spectra for a radical containing a coupled ^{14}N nucleus. **(a)** liquid-phase spectrum showing isotropic $+1$, 0 and -1 features. The remainder are simulated frozen-solution spectra ('powder' spectra); **(b)** with anisotropy in g but not in A; **(c)** with anisotropy in A but not in g; **(d)** the correct spectrum with anisotropic g- and A (^{14}N) components.

B

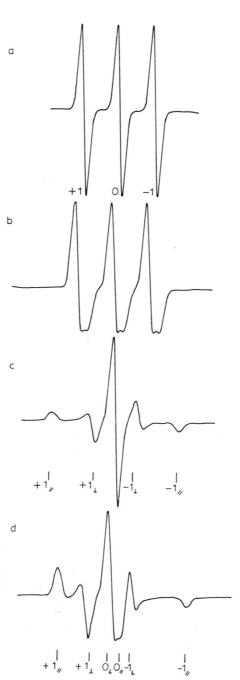

crystals), and hence obtain spectra which are generally very an-isotropic. Although the features are broadened, this effect is usually small compared with the shift, and spectra can be interpreted. In contrast, NMR spectra of frozen specimens are generally too broad for detection with standard spectrometers, the broadening being such that most features are lost anyway. (These can, of course, be extracted using magic-angle spinning (MAS) NMR methods.)

Typical solid-state spectra are shown in Fig. 3.7, together with the corresponding isotropic (liquid-phase) spectrum (*a*). This is a simple 1:1:1 spectrum for a nitrogen-centred radical such as a nitroxide. (^{14}N has $I = 1$, giving $+1$, 0 and -1 components.) In Fig. 3.7b only the *g*-anisotropy has been included (with $g_{\parallel} = 2.001$ and $g_{\perp} = 2.007$) and in Fig. 3.7c only the hyperfine anisotropy is shown ($A_{\parallel} = 45$ G, $A_{\perp} = 15$ G). These values have been selected purely for illustrative purposes. In Fig. 3.7d, the proper solid-state spectrum, including both anisotropies, is shown. Note how the combined anisotropies constrain the ($+1$) parallel and perpendicular features together, but separate out the corresponding (-1) features. From this it is easy to see why, in the liquid-phase, if the rate of tumbling is relatively slow, the widths of the features are $-1 > +1 > 0$.

3.1.6. Obtaining and interpreting spectra

As stressed above, good ESR manuals contain most of the information required in order to obtain good spectra. Here we list a few points that are somtimes overlooked.

(a) *Markers* Though not necessary, it may help to use a marker as in NMR spectroscopy. This can also serve as an intensity (sensitivity)

Fig. 3.7. First derivative liquid-phase ESR spectra for nitroxide radicals formed by the spin-trapping method. **(a)** the $\cdot CCl_3$ adduct of PBN, showing hyperfine coupling to the ^{14}N and ^{1}H of the parent nitrone; **(b)** the same species formed from $^{13}\cdot CCl_3$ radicals, showing the extra large doublet splitting from $^{13}C\cdot$ (The β-lines are from an extraneous radical species.) **(c)** the hydroxyl radical adduct of DMPO, showing the characteristic 1:2:2:1 quartet, generated because of the fortuitous equality of the ^{1}H and ^{14}N splittings.

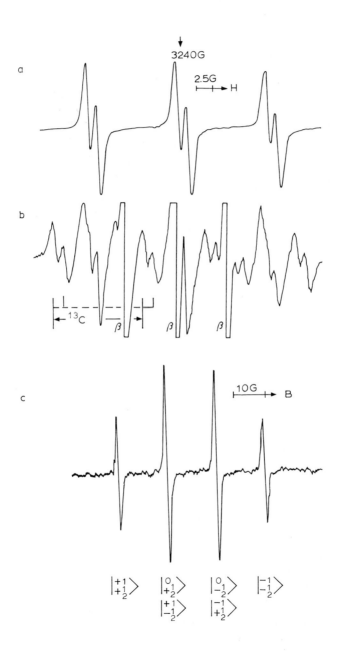

check in quantitative studies. In our view either Mn^{2+} in MgO or Cr^{3+} in ruby serve this purpose well. These can be located in the resonator well separated from the sample tube and can be raised or lowered to give weak or strong signals, as required. The Mn^{2+} marker gives six major lines separated by ca. 100 G, centred close to $g = 2$, so that the central 100 G is free of signals. The Cr^{3+} signal comes in the region of $g = 1.9$ (the ruby crystal must be held in a fixed orientation), and has four weak satellite lines from ^{53}Cr nuclei. (The major isotopes of Cr are non-magnetic.) Again the signals are usually well-removed from those for organic radicals. Others use a dual cavity system so that any marker can be used, the signal being switched in or out as required.

(b) *Power levels*   Signal intensities increase with the square root of the microwave power until saturation begins to occur. In general, the greater the anisotropy of the signal, the higher the power that can be used. Saturation is best avoided, but if a spectrum contains two components, one readily saturated and the other not, then high power will reveal the latter relatively free of the former, whilst the reverse will tend to occur at very low powers. So, vary the power to suit the spectra.

(c) *Field modulation*   The same maxim applies; use a maximum but do not overmodulate or resolution will be lost.

(d) *Sweep time*   If spectral accumulation is used, this can be as fast as you like. If not, then very long times are best. The noise then tends to combine to give, in effect, a thick line in which small changes can easily be seen.

(e) *Simulation*   When in doubt, simulate. For high-resolution liquid-phase spectra this is often unnecessary, but it becomes increasingly useful when lines overlap extensively. For anisotropic solid-state spectra, simulation is often necessary, and it may take a long time, using guessed parameters, to obtain a good fit. Beware of situations where *g*- and *A*-components may not share the same axes. There is no problem when single crystalline samples are being studied, but powder spectra can be very misleading. Programmes are available for most situations, and all we can say is that if the simple assumption of coaxial components always gives a poor fit, then try varying relative angles. Generally, one axis is common, but even so, this may require lots of computing time! Intelligent guesses may help a lot. So also will the use of Q- or S-band spectrometers, together with the X-band instrument.

(f) *Wing-features*   Always study the spectra at low and high fields outside the region of strong absorption. These regions may contain weak satellite features, or features from extra radicals that you may not have been expecting. Remember that species with broad lines spread over a wide field range are expected to give relatively weak features that can easily be missed. So use a large sweep and increase the gain, the modulation and the power to see what is present.

(g) *Resolution enhancement*   If a feature contains unresolved splittings, one can obtain remarkably improved resolution by using higher derivatives. There are convenient programmes available that do this and then translate the spectra back into first derivative format but retaining the enhanced resolution.

(h) *Corrections to data*   When hyperfine splittings are large, ($>$ ca. 100 G at X-band) the separations will not give the correct *A*-values. This is because of a zero-field coupling between electrons and nuclei, commonly known as the Breit-Rabi effect. Again, computer programmes will correct for this effect, which will not normally be encountered in biological free radical studies.

(i) *Spin-spin broadening*   This can result if the radical concentration is too high ($>$ ca. 10^{-3} M), so keep the sample dilute. Also, it can be caused by magnetic interactions with other paramagnetic material such as oxygen or transition metal ions. Remove these if you can. This can be important in liquid-phase spectra. It can be even more important in frozen solution studies if phase-separation occurs. For example, a 10^{-6} M aqueous solution of a nitroxide radical ($R_2N\cdot O$) should give a highly resolved 1:1:1 triplet (^{14}N) (possibly with further ^1H splitting). If this is frozen, only a *single* broad line will normally be observed. The same radicals in methanol, on freezing, will give a well-defined anisotropic triplet (Fig. 3.8). The problem is phase separation. Water loves water almost exclusively and, on freezing, ice crystals grow out, leaving a second phase which is highly concentrated in nitroxide. Thus, not only does spin-spin broadening occur, but spin-exchange may be so fast that the ^{14}N hyperfine coupling is lost. (Individual electrons visit many ^{14}N nuclei in the ESR time-scale and hence no defined coupling can be detected.)

3.2. Spin-trapping

Although ESR spectroscopy can be used in principle to detect any radical intermediates, some give such broad lines in the liquid phase that they are undetectable. Others may be so reactive that they never accumulate to gain detectable concentrations. In such circumstances it may be helpful to intercept radicals using molecules to which they will add to give stable adducts that can then accumulate and be studied by ESR spectroscopy with no difficulty. Almost all such 'spin-traps' give nitroxide radicals as adducts, because of the high stability of these species, and their characteristic triplet spectra for coupling to ^{14}N (Perkins, 1980; Holtzman, 1984; Mason and Mottley, 1987).

3.2.1. Spin-trapping in vitro

Examples of such reactions in chemical systems together with the symbols used, are given in Table 3.1, and typical adduct spectra are shown in Fig. 3.7. Although a number of other stable radicals could be used in this way, in practice only the nitroxide precursors are important.

Not all radicals will add to spin traps, and in some cases addition may be too slow to compete with other processes. Also in biological systems, it must be borne in mind that the spin-traps will divert the reactive radicals from their normal role. In fact, they act as radical scavengers. Scavengers are often used as probes of radical reactions, so it is useful to use spin-traps for this purpose as well as for detection by ESR spectroscopy.

If traps have to be used, identification of the parent radicals obviously becomes more difficult than when these can be detected directly. Thus the discovery of the ideal spin-trap depends in part on the resulting spectrum and its uniqueness for the trapped radical. Ideally there is extra hyperfine coupling characteristic of the trapped species. This is especially convincing if coupling to an enriched isotopic nucleus can be seen, since it establishes the involvement of the enriched specimen in a direct manner (e.g., the use of $^{13}CCl_4$ and $^{17}O_2$ discussed below). If such characteristic features are not resolved in the normal spectrum this may be achieved using resolution enhancement, or better, ENDOR spectroscopy.

If extra splittings characteristic of R$^\bullet$ rather than the spin-trap are not detected, then one relies on the measured ^{14}N coupling and any ^1H coupling from the trap. Of the traps normally used (Table 3.1) one Me_3CNO (NTB) adds the radical directly to nitrogen [3.1]. All the others are nitrones, the radical adding to carbon as in [3.2], (for DMPO). In general the nitrones are

$$R^\bullet + Me_3CNO \longrightarrow \underset{R}{\overset{Me_3C}{\diagdown}} N^\bullet\text{-}O$$

[3.1]

[3.2]

TABLE 3.1.

Some spin-traps and radical adducts.

Me$_3$C–NO   NTB   R$^{•}$ + Me$_3$C–NO ⟶

$$\underset{Me_3C}{\overset{R}{\diagdown}}N{-}O$$

$$\langle O \rangle{-}CH{=}\overset{\overset{O}{|}}{N}{-}Me_3 \quad PBN \quad R^{•} + PhCH{=}\overset{\overset{O}{|}}{N}{-}Me_3 \quad \longrightarrow \quad \underset{Ph}{\overset{R}{\diagdown}}\!\!\!\overset{|}{C}{-}\overset{\overset{O}{|}}{N}{-}CMe_3$$

$$\underset{Me}{\overset{Me}{\diagdown}}\!\!\!\underset{\underset{O}{|}}{N}\!\!{-}H \qquad DMPO$$

$$\underset{Me}{\overset{Me}{\diagdown}}\!\!\!\underset{\underset{O}{|}}{N}\!\!{-}Me \qquad TMPO$$

$$N\langle O \rangle{-}CH{=}\overset{\overset{O}{|}}{N}{-}Me_3 \qquad 4\text{-PyBN}$$

$$O{-}N\langle O \rangle{-}CH{=}\overset{\overset{O}{|}}{N}{-}Me_3 \qquad 4\text{-POBN}$$

$$Me{-}\overset{+}{N}\langle O \rangle{-}CH{=}\overset{\overset{O}{|}}{N}{-}Me_3 \qquad 4\text{-MePyBN}$$

preferred, despite the fact that there is often a greater chance of detecting specific nuclei in the R-group using NTB. One area in which NTB is most useful is when R = RO$^{\bullet}$, since the product, Me_3C-N$^{\bullet}$(O)OR' is now a nitro radical. These are pyramidal at nitrogen and the ^{14}N coupling has increased from the value of ca. 16 G, normal for nitroxides, to nearly 30 G, typical of nitroradicals. The main advantages of the nitrone traps is that the β-hydrogen, circled in equation [3.2], gives rise to a variable doublet splitting in the ESR spectra which is governed by the size and chemical nature of the R-group. It is generally more variable than the ^{14}N coupling, and hence A (^{1}H) can be used as a sensitive method of fingerprinting R$^{\bullet}$, provided the adduct has been properly characterized. A particularly important example of this is that of the hydroxyl radical adduct of DMPO. Here the ^{1}H coupling is fortuitously equal to the ^{14}N so a 1:2:2:1 quartet results which uniquely identifies the $^{\bullet}$OH-adduct (Fig. 3.7).

3.2.2. Spin trapping in vivo

For in vivo work, the nitrone spin traps have been favoured. Thus, in general, only the ^{14}N and ^{1}H coupling constants of the parent trap are measured and these are matched against reference compounds.

There are of course many problems. One is that metabolism of the trap may occur before it has captured the radicals of interest. Another is that the trap itself may upset the delicate balance of reactions which serve to produce radicals in its absence. It seems that for most traps, toxicity is not a serious problem.

One of the first in vivo studies was that of Lai et al. (1979) using the PBN trap together with tetrachloromethane fed to a rat via a stomach tube. After 2 h the liver was extracted with a methanol/chloroform mixture, and the chloroform layer was studied. The spectrum was identified as that of the $^{\bullet}CCl_3$ radical adduct. This identification was fully proven using $^{13}CCl_4$ (Albano et al., 1982; Tomasi et al., 1985). This remains one of the most successful studies of this type, but in situ detection not involving sacrifice and extraction would be most desirable.

Other similar halogenated radicals have been detected in this way, and it seems that the high toxicity of such compounds is due to radical formation probably via electron capture. The great advantage of extraction in non-aqueous solvents such as toluene is that much larger volumes can be used than with aqueous extracts. However, this discriminates against water-soluble nitroxides, which may be missed entirely.

Ron Mason and his co-workers have greatly improved this technique by using an improved (TM_{110}) ESR cavity which enables them to use 17-mm flat cells containing ca. 100 μl of fluid. Because of the low solubility of oxygen in water, deoxygenation is not necessary, and radical metabolites can be detected in urine, blood and bile fluids with no difficulty (LaCagnin et al., 1988).

For example, with urine, rats administered with $^{13}CCl_4$ and PBN gave a nitroxide radical which proved to be the $^{13\cdot}CO_2{}^-$ adduct, clearly formed by hydrolysis of $^{13\cdot}CCl_3$ radicals or their adducts (Connor et al., 1986). Similar products ($\cdot CCl_3$ and $\cdot CO_2{}^-$) were obtained from bile samples.

The reaction of oxyhaemoglobin with phenylhydrazine was studied in whole blood samples (Maples et al., 1988). Hydrazine-based drugs induce destruction of red blood cells with resulting haemolytic anaemia. Using DMPO as a trap, nitroxide radicals were detected, but these had solid-state or immobilized spectra with broad parallel and perpendicular features. Whilst no firm identification is possibly based on ESR spectra, except that the trapped radical must be a high polymer, various lines of evidence lead to the conclusion that the adduct was formed from a sulphydryl radical on oxyhaemoglobin. Chloroform extracts gave the phenyl radical adduct in accord with in vitro studies.

These are just a few examples. Spin-traps are now very widely used in biological studies, and, in general, are very effective. Since many such studies are directed at detection of $O_2{}^{\cdot-}$ and $\cdot OH$ radicals, it has become normal practice to check that the $O_2{}^{\cdot-}$ radical adduct is not formed when superoxide dismutase is added, and that the $\cdot OH$ radical adduct is not formed when catalase is added to remove H_2O_2, which

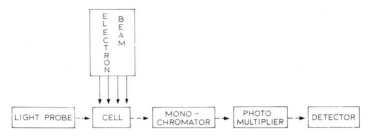

Fig. 3.8. Possible lay-out for a simple pulse radiolysis instrument. Electrons from the accelerator impinge on the cell in a short pulse which is followed at controlled time intervals by the light pulse which probes the spectra of transients and their time dependence.

is the usual source of ·OH radicals. Also, if a compound such as ethanol can be added, which can react with ·OH radicals to give a new radical with its own characteristic adduct spectrum, then replacement of the ·OH-adduct spectrum by the new one is good evidence for the involvement of hydroxyl radicals. One of the most convincing experiments in studies of oxygen toxicity is to use $^{17}O_2$. Whilst normal oxygen gives no hyperfine coupling, ^{17}O with $I = 5/2$, gives six lines, and the appearance of this extra splitting in the ESR spectra of oxygen centred adducts such as DMPO-^{17}OH provides very clear evidence of the involvement of oxygen in radical formation.

3.3. Pulse radiolysis

The importance of this technique to chemistry and biology has been far less widely accepted than it deserves. The essential technical problem involves the generation of short pulses of ionizing radiation followed generally by optical detection of transient species (Swallow, 1973; von Sonntag, 1987; Kiefer, 1990).

The basis of the method is shown in Fig. 3.8. Pulsed radiation is usually generated by linear electron accelerators (Linacs) or by Van de Graaf accelerators, each having certain advantages. Both give nanosecond at least resolution, but can be adapted to give picosecond resolution by a variety of (expensive) modifications. For most work in the biological area, the basic instruments are quite satisfactory. Whilst optical (UV) detection remains the standard procedure, there are at least two limitations. One is that UV spectra can be broad, and difficult to assign, and the other is that opaque samples are difficult to handle. The former has been considerably overcome by the great range of experience built up over many years for this and the releated technique of flash photolysis. The latter can be overcome at a qualitative level by using diffuse reflectance methods (Fig. 3.9). However, several other methods of radical detection have been used effectively, namely conductivity, ESR spectroscopy, fluorescence spectroscopy and optically

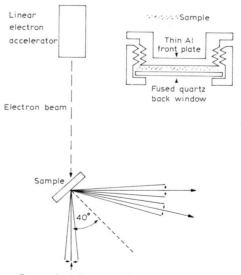

Fig. 3.9. Method used to study the diffuse reflectance of opaque samples in pulse radiolysis (Adams et al., 1991). The insert shows the method used to mount the sample. The electrons pass through the thin aluminium plate and the probe light through the quartz window. The scattered beam is analysed in the normal way with a monochromator, care being taken to exclude normal reflected light.

detected magnetic resonance. We stress that the optical studies detect any intermediate species having an intense UV-visible chromophore, be it a radical or unstable molecule or ion, conductivity detects ionformation, but ESR and ODMR detect radicals and hence are of particular interest for our purposes.

3.3.1. Optical detection of radicals

Because of the high sensitivity required, this generally involves the measurement of intense electronic transitions. In principle, infrared or Raman detection could be more widely applicable, but, except for resonance Raman methods, which again depend on the presence of UV-visible bands, these methods are too insensitive to be of much use at present.

Thus, the species most widely studied are those with low-lying excited states. These include proteins containing aromatic side-chains such as tyrosine, tryptophan or imidazole side groups and in particular disulphide derivatives, and metallo proteins especially haem derivatives. Also nucleic acid bases have been widely studied. These studies have been extended to cover dimers and oligomers, and various forms of DNA itself (Chapter 8). Other biological studies using this technique include photosynthesis, electron-transport systems, a wide range of drugs and the process of vision. Examples of the use of pulse radiolysis with optical detection are given throughout this work. One specific example is described here to illustrate the method.

Haem proteins   In almost all pulse-radiolysis studies, dilute aqueous solutions are used, so direct damage is unimportant. It is clear that e^-_{aq} can add to oxidized metal centres. Thus, for example, metmyoglobin gives ferro- (or deoxy)myoglobin, the electron being fairly specific in its reaction (Whitburn et al., 1982). The reaction can be followed by studying the loss of the strong visible band and the change in the haem iron bands. However, reactions of hydroxyl radicals are far from clear, although some reduction to the ferro- form still seems to occur, which is not easy to understand. We would expect fairly indiscriminate attack on the protein unit to give a range of radical centres, some of which might end up by inducing electron donation to Fe(III). We stress, however, that this product is unlikely to be pure ferromyoglobin. The problem is that the optical spectra for Fe(II) haem units in a myoglobin molecule that has been modified at sites other than the haem unit will probably be indistinguishable from that for the pure ferro-derivative.

3.3.2. Conductivity detection of radicals

As shown in equation [2.13], radiolysis of aqueous solutions results initially in the formation of H_3O^+ in high yield. This can be studied by measuring changes in the conductivity, and this constitutes a powerful method of studying reaction rates. The method hinges strongly on the fact that the mobilities of H_3O^+ (and OH^-) ions are very high relative to all other ions, and hence changes in their concentrations tend to dominate the results. These studies are generally complementary to optical studies. However, in the presence of buffers, results are far more difficult to utilize (Goebl, 1984). Hence for most biological systems the technique is less widely used.

One interesting example in which changes in $[H_3O^+]$ or $[OH^-]$ are not directly of interest is the study of strand-breaks in polynucleotide chains (Hildenbrand and Schulte-Frohlinde, 1989). This study follows changes in the concentration of 'free' K^+ ions using conductivity. Because of the huge negative charge on DNA strands, there is extensive 'ion-pairing', the paired K^+ ions making no significant contribution to the conductivity. When strand breaks occur there is an increase in the number of 'free' K^+ ions and hence the conductivity increases. These conductivity techniques have been reviewed by Asmus and Janata (1982). The time-scales cover the range from ca. 50 ps down to ca. 100 ns.

3.3.3. Detection of radicals by Raman and resonance Raman spectroscopy and light scattering

Resonance Raman spectroscopy is a sensitive technique which can complement normal optical detection of intermediates. It is currently being exploited particularly by Schuler and his co-workers (see, for example Qin et al., 1985). The technique of light scattering is less directly useful, especially for biological systems. Rayleigh scattering increases with the molecular weight of the scattering molecules. Hence it can be used to study chain breaks and cross-links in DNA, although interpretation is difficult. Key references are Beck (1979) and Washino and Schnabel (1982).

3.3.4. Detection by ESR spectroscopy

Although, in principle, ESR spectroscopy is the most powerful method for detecting radical intermediates, it has not been widely used in conjunction with steady-state or pulse radiolysis, mainly because of technical difficulties. Important early work was done by Fessenden, Schuler and their co-workers (see for example Eiben and Fessenden, 1968; and Fessenden and Schuler, 1971) using steady-state radiation from a Van der Graaf accelerator. In this way, liquid-phase ESR spectra were generated for a range of radicals never previously observed by ESR methods.

It is necessary to focus the electron beam through a hole in the electromagnet so that the beam is parallel to the magnetic field. The beam passes into the cavity, generating radicals in situ. Whilst ideal for detecting and identifying radicals with multiple narrow-line spectra, the method is best carried out together with optical studies, since *g*-values and reaction rates are less readily estimated from the ESR spectra. With modern technology it is possible to link a pulsed ESR spectrometer to a pulse-radiolysis system. As stressed in Section 3.1, one great advantage over other detection methods is that the sample need not be transparent. This means that cellular and tissue specimens can be studied.

3.3.5. The use of pulse-ESR and Fourier Transform Techniques

Now that commercial pulse-ESR spectrometers are available, there can be no doubt that this will become a popular instrument, especially for those interested in studies in the time domain. At present, there are no great advantages in the sensitivity, but there probably will be in the near future. Work in this area is covered in two recent books (Keran and Bowman, 1990; Holt, 1989). An important aspect of pulsed ESR techniques is that they can be used to measure spin-lattice relaxation rates specifically. These are generally not obtained from normal CW-ESR spectra, which are frequently insensitive to this parameter.

Another major development in the field of pulsed ESR is the use of Fourier-transform methods. It is this combination that has led to major improvements in NMR spectroscopy; however, unfortunately, it is still of relatively limited use to ESR spectroscopists because of the difficulties involved in covering a wide enough range of field. In this technique, the high-resolution spectrum is recovered by Fourier transformation of the entire time-domain signal after one or several pulses (Angerhofer et al., 1988).

3.3.6. Detection using fluorescence-detected magnetic resonance (FDMR)

This powerful technique developed in particular by Trifunac and co-workers (Smith and Trifumac, 1981; Jonah, 1988) depends on the use of a fluorescing 'marker' which indirectly detects the presence of radical ions.

3.4. Chemical measurement of the hydroxyl radical

To assess the damaging role of hydroxyl radical generation in in vitro systems it is obviously important to be able to detect, identify and quantitate it accurately. Since hydroxyl radical generation in vivo has also been suggested to be important, a method that could be applied in vivo would be of great value. Currently, no such method exists, although there are interesting developments on the horizon, as this chapter will outline.

Direct measurement of hydroxyl radical production by cells is technically rather difficult. Some of the most commonly used methods for the measurement of the hydroxyl radical (Table 3.2) require technologies which are expensive and not normally readily available. Among these are:

1. The decarboxylation of radio-labelled benzoic acid, the labelled

TABLE 3.2.
Summary of some of the literature studies pertaining to the measurement of hydroxyl radical production (from Greenwald et al., 1989).

Substrate	Product	Technique	Reference
Methional	ethylene	gas chromatography	Tauber and Babior, 1977
Methional	ethylene	gas chromatography	Klebanoff and Rosen, 1978
KMB	ethylene	gas chromatography	Weiss et al., 1978
KMB	ethylene	gas chromatography	Ambrusco and Johnston, 1981
KMB	ethylene	gas chromatography	Janco and English, 1983
KMB	ethylene	gas chromatography	Niwa et al., 1984
KMB	ethylene	gas chromatography	Miyachi et al., 1987
DMPO	DMPO/·OH	ESR	Britigan et al., 1986
			Green et al., 1979
			Rosen and Klebanoff, 1979
DMPO	DMPO/·OOH	ESR	Bannister et al., 1982
DMSO	methane	mass spectroscopy	Repine et al., 1979
Benzoic acid	$^{14}CO_2$	scintillation counter	Sagone et al., 1980
Benzoic acid	$^{14}CO_2$	scintillation counter	Green et al., 1985
Phe	Tyr isomers	HPLC	Fujimoto et al., 1987

Abbreviations: KMB, 2-keto-methylthiobutyric acid; DMPO, the spintrap, 5,5-dimethyl-1-pyroline-N-oxide; ESR, electron-spin resonance (= electron paramagnetic resonance, EPR) spectroscopy; DMSO, dimethyl sulphoxide; Phe, phenylalanine; Tyr, tyrosine; HPLC, high-performance liquid chromatography.

CO_2 being detected in a special apparatus by gas-trapping agents and quantified by scintillation counting;

2. reaction with the spin trap 5,5-dimethyl-1-pyrroline-N-oxide (DMPO) to produce an adduct detectable by electron paramagnetic resonance (see Sections 3.1 and 3.2); and

3. the conversion of methional or 2-keto-4-methylthiobutyric acid (KMB) to ethylene, which is detected by gas chromatography-mass spectroscopy (see Chapter 7).

These methods are not considered in detail herein. Two simpler tests that are widely used for hydroxyl radical production are described in Sections 3.4.1 and 3.4.2.

3.4.1. Deoxyribose assay for the detection of hydroxyl radical production in cellular systems

Deoxyribose reacts with hydroxyl radical with a rate constant of 3.1×10^9 M^{-1}·s^{-1}. Deoxyribose degraded by hydroxyl radical produces malonyldialdehyde which reacts with thiobarbituric acid (TBA) to produce a pink chromophore absorbing at 532 nm (Halliwell and Gutteridge, 1981).

Whether 'free' MDA is formed from the deoxyribose sugar or whether the ring is cleaved to give a product that, upon heating with TBA, forms MDA, is not clear.

This simple reaction has been exploited extensively in several cell-free systems for studying oxygen-radical reactions. The deoxyribose assay is a sensitive assay for hydroxyl radicals. However, it may not be specific for this species, and it has been suggested that deoxyribose can be degraded by ferryl haem species, although we have found no such interaction in our conditions.

This assay has recently been applied (Greenwald et al., 1989) to the detection of hydroxyl radicals formed from superoxide radical generated by stimulated polymorphonuclear leukocytes in the presence of exogenous chelated iron. The method requires only a spectrophotometer and a shaking waterbath and is therefore much more readily applied in standard biochemical research laboratories in a variety of pharmacological and pathological studies on the role of oxygen-derived free radicals in various processes. Full details of this assay are given below.

Reagents/assay tube
1. 0.5 ml (20 mM) deoxyribose;
2. 0.01 ml (10 mM) iron(III)-EDTA: MUST BE FRESHLY PREPARED; 2.7 mg iron(III) chloride in 1 ml 20 mM EDTA, prepared in water, adjusted to pH 7.4;
3. 0.28 ml of Dulbecco's phosphate-buffered saline: prepared from a 10-fold concentration stock followed by addition of 1 mg/ml glucose and adjustment to pH 7.4 with sodium hydroxide;
4. 0.2 ml suspension of polymorphonuclear leukocytes (2×10^6 cells);
5. 0.01 ml fresh phorbolmyristate acetate (PMA) solution for activating the neutrophils: $1 \rightarrow 100$ dilution on stock of 1 mg/ml in 100% dimethyl sulphoxide;
6. 0.5 ml thiobarbituric acid prepared as 1% (w/v) in 0.1 N sodium hydroxide;
7. 0.5 ml trichloroacetic acid prepared as 2.8% (w/v) in water.

Thus, in 1 ml of reaction mixture the final composition is:
10 mM deoxyribose
0.1 mM iron(III)-EDTA
2×10^6 cells
50 ng PMA (final concentration of DMSO 0.7 mM).

Procedure
1. The reagents in phosphate-buffered saline are added, in the order stated above, to 2 ml polypropylene micro (Eppendorf) centrifuge tubes.
2. The reaction is initiated by adding 0.01 ml of fresh phorbolmyristate acetate (PMA) solution (1:100 dilution of 1 mg/ml in phosphate-buffered saline).
3. The tubes are capped and incubated for the requisite time period (recommended 30–45 min) in a shaking water bath at 37°C.

4. Blanks without PMA are incubated in parallel.
5. The reaction is stopped by rapid centrifugation at high speed in a microfuge for 2–5 min.
6. The degradation of deoxyribose is determined by a method akin to the conventional thiobarbituric acid assay for aldehydic products (see Chapter 5 for more information).
(a) 0.8 ml of each the experimental tubes and the blanks is transferred to a glass tube after the reaction is stopped;
(b) 0.5 ml of thiobarbituric acid is added, the solution vortexed and 0.5 ml trichloroacetic acid added and the solution mixed thoroughly.
(c) The tubes are heated at 100°C for 15–20 min and the absorbance measured at 532 nm.
   (Alternatively the chromogen can be extracted into butan-1-ol and the upper layer examined by measuring the fluorescence at 553 nm on excitation at 532 nm.)

The standard curve is set up from malonyldialdehyde prepared from freshly hydrolysed malonyldialdehydebisacetal: the standard (1 g/ml) is diluted 1:1000 in phosphate-buffered saline and diluted again 1:1000 to a working concentration of 1 ng/μl. Tubes containing a range of aliquots from the diluted stock, ranging from 100 μl up to 500 μl, were brought up to a volume of 0.8 ml and then the thiobarbituric acid and the trichloroacetic acid added as described above for the samples.

Typical absorbances at 532 nm are: 0.054 for 100 ng MDA, and 0.268 for 500 ng MDA.

Precautions
1. The isotonicity must be carefully maintained in all the solutions added to the reaction mixture.
2. The ferric ion needs to be chelated to be effective, and EDTA was found to be the most potent from a range of chelators tested (Greenwald et al., 1989). Cupric ion should not be used as it is a poor catalyst of the reaction.
3. Iron levels at 0.1 mM are optimal; above 0.2 mM a pale orange

cólour appears in the blanks and the control tubes, thus interfering with the accuracy of the readings in the assay.
4. The deep pink colour of the tissue medium must be totally washed off.

Essential controls
The following checks must be made when examining biological systems by this method for the generation of hydroxyl radicals:
1. The material under examination should not react rapidly with hydrogen peroxide, which could block hydroxyl radical formation (this is rarely a problem as hydrogen peroxide is relatively unreactive).
2. The material under examination must not be a powerful iron chelator, capable of removing iron ions from EDTA. This is again rarely a problem.
3. Attack of hydroxyl radical on the material under investigation should not produce thiobarbituric acid-reactive material; hence a control is performed in which deoxyribose is omitted from the reaction mixture.
4. The system being assessed should not interfere with the measurement of the deoxyribose degradation products. This can be checked by showing that it does not inhibit when added to the reaction mixture at the end of the incubation with the thiobarbituric acid and the acid.

3.4.2. Hydroxylation of salicylate: measurement of 2,3-dihydroxybenzoate

Principle
Halliwell (1978) proposed aromatic hydroxylation as an assay for ·OH radical production. Hydroxylated products are quantitated using a colorimetric method that measures *o*-dihyric phenols. The assay has since been improved by Richmond et al. (1981) and has often been used as a simple 'test tube' assay of ·OH formation (Halliwell et al., 1988) in activated phagoctic cells, isolated hepatocytes, soluble en-

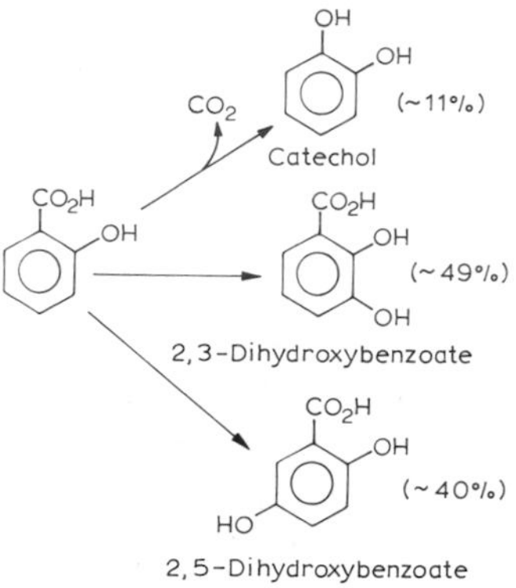

Fig. 3.10. Products of hydroxyl radical attack on salicylate.

zymes, reperfused post-ischaemic tissues and similar biological systems. Aspirin administered to humans or to animals is rapidly hydrolysed to salicylate. Attack of hydroxyl radicals generated upon salicylate at pH 7.4 results in three products (Fig. 3.10).

The major products are 2,3- and 2,5-hydroxybenzoate, but a small amount of catechol is formed by decarboxylation. The dihydroxy benzoates are stable in air in body fluids and tissue extracts, provided that the pH is not allowed to rise.

Thus the hydroxylation of salicylate can be used in vivo to assess ·OH radical formation. These products have in fact been detected in trace amounts in urine and blood plasma from human volunteers after consumption of aspirin (Grootveld and Halliwell, 1986, 1988).

We describe here the method for detection of hydroxyl radical formed from an enzymic superoxide-generating system in the presence of a low-molecular-weight iron-chelate.

Reagent preparation
The reaction mixture contains the following reagents (hypoxanthine solutions are made up initially in 50 mM NaOH and diluted with KH_2PO_4/KOH buffer, pH 7.4, 150 mM in phosphate, to give the concentrations required; salicylate is dissolved in the same buffer):

40 μl $FeCl_3$, 5 mM (prepare fresh before use)

40 μl EDTA, 5 mM
200 μl hypoxanthine, 2 mM
200 μl salicylate, 20 mM
1.48 ml KH_2PO_4 buffer

Procedure

1. The reaction is started by adding 40 μl of xanthine oxidase (diluted just before use into buffer to give 0.4 enzyme units/ml) to the 2 ml of reaction mixture.
2. The tubes incubated with gentle shaking at 25°C for 90 min.
3. The reaction is stopped by adding 80 μl of 11.6 M HCl and 0.5 g of solid NaCl followed by 4 ml of chilled diethyl ether (ethoxyethane) and vortexed for 30 s.
4. 3 ml of the upper (ether) layer is removed carefully using a syringe and evaporated to dryness at 40°C in a boiling tube in a fume cupboard.
5. The residue is dissolved in 0.25 ml of double-distilled water and the following reagents are added in the order stated:
   125 μl 10% (w/v) trichloracetic acid dissolved in 0.5 M HCl,
   0.25 ml 10% (w/v) aqueous sodium tungstate,
   0.25 ml 0.5% (w/v) aqueous sodium nitrite (made up fresh daily), and reagents are mixed well.
6. Allow to stand for 5 min.
7. 0.5 ml of 0.5 M KOH is added, and the absorbance of the pink complex read at 510 nm after 60 s.
8. Standard curves are prepared using 2,3-dihydroxybenzoate carried through the same ether extraction and colorimetric assay.

A complete hypoxanthine–xanthine oxidase system should give a final A_{510} of about 0.65, corresponding to 150–200 nmol of hydroxylated product. Formation of hydroxylated products can be inhibited almost completely by superoxide dismutase, catalase, or the iron chelator desferrioxamine (Richmond et al., 1981).

This method gives, if anything, an underestimate (only 50–70%) of hydroxylated products, and thus of hydroxyl radical production, since only a single hydroxylated product is being measured applying this spectrophotometric assay.

For a more quantitative approach, HPLC detection is recommended (Halliwell, 1988). However, the spectrophotometric assay described here is certainly ideal for qualitative assessment of hydroxyl radical production.

Fig. 3.11 shows chromatograms of 2,3-dihydroxybenzoate and 2,5-dihyroxybenzoate separated by HPLC reversed-phase chromatography fitted with electrochemical detection (Halliwell et al., 1988). HPLC is carried out on a Spherisorb 5ODS column (25 cm × 4.6 mm) applying 97.2% (v/v) 30 mM sodium citrate/2.7 mM acetate

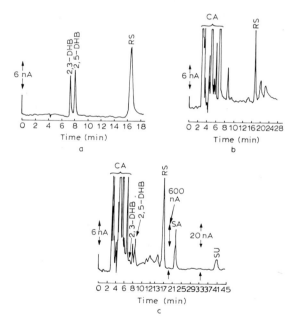

Fig. 3.11. HPLC of the hydroxylated products of salicylate (from Halliwell et al., 1988): 2,3-DHB, 2,3-dihydroxybenzoate; 2,5-DHB, 2,5-dihydroxybenzoate; RS, resorcinol; SA, salicylate; SU, salicylurate (a major metabolite of aspirin); CA, catecholamines. (a) Separation of a standard mixture of 2,3-dihydroxybenzoate, 2,5-dihydroxybenzoate and resorcinol (internal standard); (b) Separation of an extract of a plasma sample from a healthy individual not consuming aspirin; (c) Separation of an extract of a plasma sample from a healthy individual consuming aspirin (the large peaks to the left of the chromatogram are probably catecholamines – marked CA).

buffer (pH 4.75)/2.8% (v/v) methanol as eluant at a flow rate of 0.9 ml/min and at ambient temperature.

Advantages
Aromatic hydroxylation of this type is a highly specific test for hydroxyl radicals, provided the ionization potential of the aromatic compound is not too low.

Disadvantages
The most serious disadvantage of this technique is that salicylate and catechols have high affinities for iron(III). The resulting complexes can react with hydrogen peroxide to give hydroxyl radicals (via the iron(II) species) which will obviously interfere with the assay. Another problem is that these complexes absorb strongly in the 510 nm region and hence, if formed, will interfere with the assay. 2,5-Dihydroxybenzoate measurement could give misleading results. Applying the salicylate assay 2,3-dihydroxybenzoate is the main product of interest (Halliwell et al., 1991).

3.5. Chemical measurement of superoxide radical

3.5.1. The formazan reaction

3.5.1.1. The nitroblue tetrazolium reduction
Introduction   The reduction of yellow nitroblue tetrazolium (NBT) to blue formazan is applied as a probe of $O_2^{\cdot-}$ generation in biological systems. This reaction is utilized in demonstrating the role of phagocytes (neutrophils and monocytes/macrophages) in the host response to infection and inflammation: once stimulated by certain ligand-receptor interactions or particle ingestion, these cells are capable of consuming increased amounts of oxygen. The oxygen is rapidly reduced via a complex NADPH–oxidase system comprising a flavoprotein and a *b*-type cytochrome (Segal and Jones, 1979). The cytochrome has a sufficiently low midpoint potential to allow the

direct catalytic transfer of electrons from NADPH to oxygen, resulting in the production of superoxide.

The first quantitative NBT test was developed by Baehner and
Natham in 1968 to detect neutrophil dysfunction in patients with
chronic granulomatous disease (CGD). The neutrophil membranes of
CGD patients lack the specific cytochrome b_{245} which is a key component in generating superoxide by the cell, following bacterial stimulation of the cell. As a result the patients' neutrophils are incapable of
responding with a respiratory burst, failing to generate the free radicals necessary for the antimicrobial function of this cell.

Principle   The respiratory burst of phagocytic cells can be assessed
by incubating a suspension of the cells in an isotonic solution of the
yellow oxidised nitroblue tetrazolium (NBT) dye. During this process,
the soluble dye interacts with the cytoplasmic components associating
with the oxidant species generated. Although the NBT test is not a
specific marker for superoxide production, NBT reduction by activat-

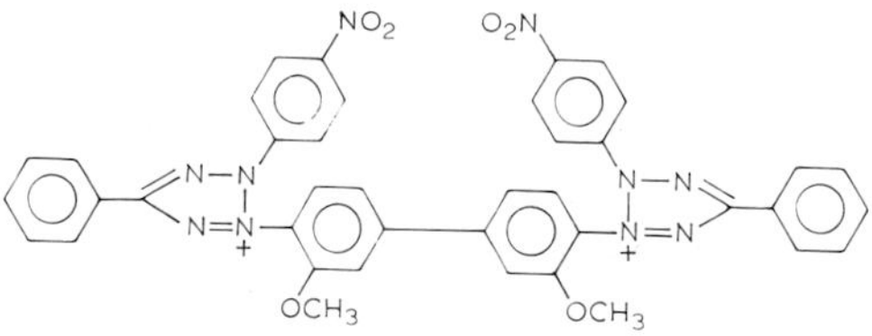

(A)

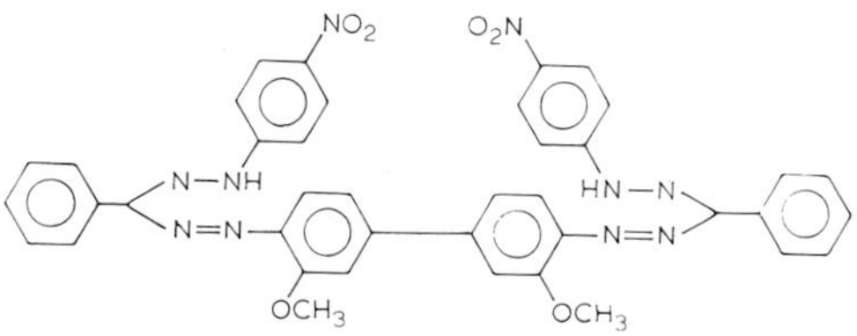

(B)

Scheme 3.1. Structures of **(A)** NBT; **(B)** formazan.

ed cells in the presence of superoxide dismutase has been shown to be markedly reduced (Maly et al., 1989), which suggests the major oxidant species responsible for the reduction of dye to a black-blue deposit called formazan is superoxide (Scheme 3.1). The overall degree of NBT reduction in a given cell population can be quantified by measuring the concentration of reduced NBT or formazan spectrophotometrically.

In normal healthy individuals, the spontaneous reduction of NBT dye by neutrophils is very low (less than 10% of the cells are positive). A low basal NBT response is an important negative control during the performance of any NBT test, as it monitors the non-stimulatory nature of the test reagents and procedure. Stimulated NBT tests are carried out as a positive control. By deliberately stimulating cells with specific stimulating agents, such as phorbol myristate acetate (PMA) one can activate cells from normal individuals to reduce NBT in a dose-responsive manner.

Factors that influence NBT reduction   There are many different versions of the 'NBT test'. A number of authors have examined the various technical details which influence the performance of the assay (Merzbach and Obedeanu, 1974; Eggleton, 1987). The percentage of cells reducing NBT can be affected by (a) prolonged sample storage; (b) use of excessive concentrations of anticoagulants; (c) use of different concentrations and commercial preparations of dye; (d) prolonged incubation at 37°C; (e) cell contact with glass; (f) contamination with endotoxin; and (g) variation in pH.

Application of NBT tests   The NBT test has been most commonly used to assess the oxidative responsiveness of neutrophils and to test the effects of various compounds on neutrophil and monocyte/macrophage function. In this way it has been shown that leukotriene B_4 (Fletcher, 1986), granulocyte-macrophage colony-stimulating factor (Fletcher and Gasson, 1988) and various polymers (Eggleton et al., 1989) can modulate the function of mature neutrophils by enhancing the proportion of cells capable of reducing NBT. In addition, an NBT assay has been used to assess macrophage activation by γ-interferon (Rook et al., 1985). Although most of the work described below refers to neutrophils, the principles of the assay applies to other phagocytic cells.

Spectrophotometric NBT assay for isolated cells   During the 1970s a number of groups developed spectrophotometric assays to assess NBT reduction (Segal and Levi, 1975). These assays required the cells to be lysed in a detergent, followed by acid washing before the formazan was finally extracted in pyridine and read on a manual spectrophotometer. With the development of microELISA plate readers, semi-automated methods have been described (Pick et al., 1981). In 1985, Rook and colleagues described a simple microELISA method by which NBT formazan could be dissolved without heating or the need to use acid or pyridine. Instead they used potassium hydroxide to lyse the cells and dimethyl sulphoxide to enhance the sensitivity of the microELISA method. The test described below is a simplified procedure based on their assay and can be used to assess NBT reduction in isolated neutrophils, monocytes and macrophages.

Procedure

EQUIPMENT

1. 5 ml plastic blood collection tubes.
2. Variable pipettes (volume to be dispensed 10, 90, 100, 125 and 400 μl).
3. 1.5 ml Eppendorf tubes.
4. Water bath set at 37°C.
5. Eppendorf microcentrifuge.
6. Titertek multiscan MC plate reader.

REAGENTS

1. NBT dye (Sigma, grade III, crystalline).
2. Sterile endotoxin free water for irrigation (Travenol).
3. Ten times and single strength HBSS (Hanks' balanced salt solution).

4. Phorbol myristate acetate (Sigma).
5. 2 M potassium hydroxide.
6. Dimethyl sulphoxide (BDH).

PREPARATION OF STOCK SOLUTIONS
0.1% (w/v) NBT dye solution with and without PMA. 20 mg is weighed into sterile universal, 18 ml sterile water is added, followed by 2 ml ten-times strength HBSS but only after the dye has fully dissolved. On the day of use, NBT dye containing 100 ng PMA/ml is also prepared which acts as a stimulated control.

METHOD (see Fig. 3.12)
1. 100 μl of cells (10^5–10^6/100 μl) are added to an eppendorf tube and prewarmed at 37°C for 5 min in a water bath. At the same time the NBT solutions are prewarmed.
2. The NBT reaction is started by mixing equal volumes of cells and NBT solutions. The NBT-incubation mixtures are incubated further at 37°C for 15 min. The tubes are then placed on ice for 5 min.

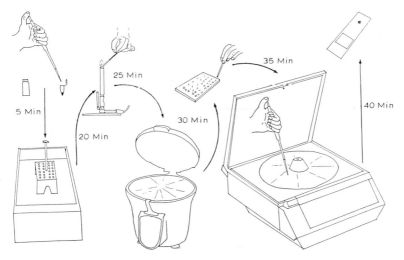

Fig. 3.12. Schematic representation of the procedure for the nitroblue tetrazolium assay (NBT).

3. The reaction mixtures are then spun in a microcentrifuge for 2 min at 10 000 rpm, and the supernatant discarded.
4. The cells are washed in 400 μl HBSS and centrifuged as in (3).
5. The resulting cell pellets are resuspended, fixed and washed in 70% (v/v) methanol for 5 min. The methanol in addition to fixing the cells removes any unreduced NBT which may have adsorbed onto the plastic. This is important as residual NBT can react with the potassium hydroxide in the next step and produce additional blue colouration.
6. The cell pellets are resuspended in 100 μl 2 M potassium hydroxide, with the aid of vigorous vortexing, the cells rapidly lyse, releasing the formazan into the potassium hydroxide.
7. 125 μl DMSO is added to each sample, which results in the development of an intense turquoise colour in positive samples. The addition of DMSO solvent to the eppendorf directly allows any residual formazan to be extracted from the plastic.
8. The complete sample is then transferred to a well of a flat bottomed microtitre plate. After 30 min the tray is placed in the plate reader and the absorbances of each sample read at 620 nm.

CALIBRATION OF THE NBT REDUCTION
The intense blue colour formed by solubilization of formazan can be directly derived from NBT, by dissolving a known quantity in potassium hydroxide and DMSO. The potassium hydroxide completely reduces the NBT present. This allows absorbance reading to be related directly to nmoles of reduced NBT/sample.

3.5.1.2. 3-(4,5-Dimethylthiazol-2-yl)-2,5-diphenyl tetrazolium reduction
A tetrazolium salt (MTT) 3-(4,5-dimethylthiazol-2-yl)-2,5-diphenyl tetrazolium bromide is able to enter cells and therefore assay levels of superoxide, without disrupting the cell with DMSO, forming the insoluble formazan product (Slater et al., 1963). Like the NBT test, MTT reduction is not specific for superoxide.

Procedure   Cell suspensions at 2×10^5/ml are maintained in Eagle's MEM supplemented with 10% (v/v) fetal calf serum. 50 μl of MTT solution (5 mg/ml in phosphate-buffered saline) are added to each well and the plates incubated at 37°C for the relevant time intervals. The medium containing MTT is then removed by aspiration very carefully and 100 μl DMSO added to each well, and the plates gently agitated until the MTT formazan has dissolved. The absorbance of each well is read with a plate reader at 570 nm, subtracting the absorbance at 750 nm as a reference (there is no absorbance by MTT at 570 nm). The plates are read within 40 min of adding DMSO.

There are disadvantages to the MTT assay   The absorbance is preferably read within 1 h, as the absorbance decreases with time; for example, 8% loss over 2 h after solubilization in DMSO. The major disadvantage is the limitation imposed by uncertainties in absorbance readings at low values.

Advantages   This assay in conjunction with the use of Multiwell plates, multichannel pipettes, microplate washer and shaker, together with a plate reading spectrophotometer, enables a large number of samples to be processed quickly. Interfacing the plate reader with a computer would provide an obvious additional advantage.

3.6. Methods of measuring hydrogen peroxide

3.6.1. Introduction

The rate of hydrogen peroxide generation in biological systems and subcellular organelles is small, of the order of nmol/min per mg protein in biomembranes (Ramasarma, 1982) and mitochondria from rat liver or from rat or pigeon heart generate 0.3 – 0.6 nmol/min per mg protein (Boveris et al., 1972; Boveris and Chance, 1973; Nohl and Hegner, 1978). The low physiological levels of hydrogen peroxide and

superoxide radical maintained by enzyme systems that actively metabolize them, together with their rapid disappearance in anoxic tissue,
render extremely difficult the direct measurement of the steady state
concentration of these 'active oxygen species' in cellular systems.
Hence methods need to be highly sensitive to detect such low levels.

3.6.2. Loss of fluorescence of scopoletin

Methods for detecting hydrogen peroxide are based on the peroxidase
assay systems. The most common employs horseradish peroxidase
(HRP) which uses hydrogen peroxide to oxidize scopoletin into a non-
fluorescent product (Loschen et al., 1971), originally described by Andreae (1955).

Certain fluorescent compounds, such as the coumarin derivative,
scopoletin, can act as hydrogen donors in the oxidative reaction catalysed by horseradish peroxidase (Udenfriend, 1969), an enzyme that
exhibits substrate specificity for hydrogen peroxide

SCOPOLETIN

(7-hydroxy-6-methoxy coumarin)

(Maehly and Chance, 1954). The rate of decrease in the fluorescence
intensity, through oxidation of scopoletin demonstrates directly the
rate of formation of hydrogen peroxide. During its oxidation by horseradish peroxidase, fluorescence is lost with a stoichiometry directly
proportional to peroxide concentration.

Thus, if a putative radical scavenger is incubated with hydrogen
peroxide and the reaction mixture sampled for analysis of hydrogen
peroxide at various times, the rates of loss of hydrogen peroxide can
be measured to allow calculation of rate constants.

The essential control is to ensure that the substance being tested is
not itself a substrate for peroxidase, which could compete with scopoletin and cause artefactual inhibition.

Principle of the detection of hydrogen peroxide by the horseradish
peroxidase-induced oxidation of reduced scopoletin

$$\text{SCOPOLETIN}_{red} + H_2O_2 \xrightarrow{\text{HRP}} \text{SCOPOLETIN}_{ox} + 2H_2O$$
$$\text{(fluorescence at 460 nm)} \qquad \text{(no fluorescence)}$$

Calibration of the system   The scopoletin concentration is adjusted in such a way that its fluorescence exceeds the substrate-induced fluorescence at least tenfold. Any marked decrease in fluorescence intensity therefore demonstrates an oxidation of scopoletin by H_2O_2 via horseradish peroxidase. Hence, after suitable calibration of the system, the rate of loss of fluoresence can be used to measure the rate of hydrogen peroxide production.

Method   Scopoletin (2–4 μM final concentration) is added to cuvette (1 cm path length) containing 2.5 ml buffer and a known concentration of cells (after removing the serum) in the range of $(1 - 2.5) \times 10^6$/ml. One of the considerations when selecting the concentrations is that the solution must be optically clear with no turbidity.

The solution is maintained at 37°C in thermostatically controlled cuvette holder. At excitation wavelength of 350 nm and the emission wavelength set at 460 nm, horseradish peroxidase is added at a final concentration of 22 nM.

The system is standardized in the absence of cells with known amounts of peroxide either generated as H_2O_2 or from glucose in the medium and glucose oxidase or added directly as ethyl peroxide. The relationship between fluorescence intensity of 2 μM scopoletin and peroxide concentration is shown in Fig. 3.13.

Limitations of the method
1. The change in fluorescence intensity produced on reaction of scopoletin with hydrogen peroxide must largely exceed the variance in fluorescence arising from, for example, intracellular metabolism.
2. The determination only covers H_2O_2 released from cells (or subcellular organelles) under investigation.
3. Hydrogen peroxide destruction from alternative pathways must be

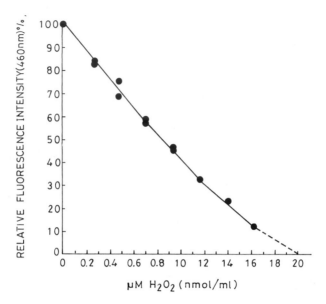

Fig. 3.13. The effects of peroxide concentration on the fluorescence of scopoletin.

negligible. Thus the intervention of intracellular or intraorganelle enzymes such as catalase or glutathione peroxidase which actively metabolize H_2O_2 may lead to an underestimate of the rate of H_2O_2 formation.

4. Competing hydrogen donors to horseradish peroxidase must be absent from the sample or perfusion medium, otherwise the decrease in scopoletin fluorescence would not represent the actual rate of hydrogen peroxide formation. A drawback of these assays is that continuous measurement of hydrogen peroxide in neutrophil lysates is not feasible because NADPH interferes with the catalytic action of the peroxidase.

5. Other substrates for the peroxidase will compete with the scopoletin, thus an underestimate of hydrogen peroxide production will result. Thus large amounts of ascorbate will interfere as will thiols.

6. Azide can be used to inhibit other haem proteins, e.g., catalase, myeloperoxidase, eosinophil peroxidase, at levels <2 mM azide

which does not interfere in the assay (Boveris et al., 1972). However, glutathione peroxidase, which also interferes in this assay (Hamer and Roos, 1985), is not inhibited by azide, or other known inhibitors.

7. The fluorescence intensity decreases at more acidic pH values.

3.6.3. Cytochrome-c peroxidase

In many systems NADPH will inevitably be present as a substrate for oxygen reduction to hydrogen peroxide. To avoid this problem Kakinuma et al. (1977) introduced the highly specific cytochrome-c peroxidase assay to measure hydrogen peroxide generation as a function of time, in this instance in leukocyte subcellular fractions (Boveris et al., 1972). The formation of a stable complex between cytochrome-c peroxidase and H_2O_2 can be measured by dual-wavelength spectrophotometry (Chance et al., 1979) or using a spectrophotometer with a multiple wavelength memory storage facility. This assay can be used in the presence of NADPH because the cytochrome-c peroxidase is highly specific for its hydrogen donor, reduced cytochrome c.

Principle of the method
Cytochrome-c peroxidase forms a stable complex with hydrogen peroxide (Yonetani, 1965; Yonetani and Ray, 1965) with an absorption maximum at 419 nm, whereas that of the free enzyme is at 407 nm.

(1)   cyt-c peroxidase + H_2O_2 →   {cyt-c peroxidase − H_2O_2}
           (407 nm)   (419 nm)

                                                    FERRYL HAEM
(2)   {cyt-c peroxidase − H_2O_2} +   2 cyt c^{2+}
                                          ↓
            cyt-c peroxidase + 2 cyt c^{3+}

Method
This procedure requires the use of a dual wavelength spectrophotometer, or a multi-wavelength computerized spectrophotometer, the

increase in absorbance at 419 nm, with an active reference wavelength at 407 nm, of reaction mixtures containing 1–2 μM cytochrome-c peroxidase and a concentration of the subcellular fraction under investigation equivalent to about 100–300 μg protein/ml.

After calibration (and equalization of light beams where relevant), the recording commences. On addition of the appropriate substrates, the decrease in absorbance at 407 nm and the simultaneous increase in $A_{419\ nm}$ (appearance of the {cytochrome-c peroxidase - H_2O_2} complex) are indicated as an increase in $A_{419-407}$. This is linear with the concentration of hydrogen peroxide consumed within the (cytochrome-c peroxidase)-H_2O_2 complex.

Advantages of the assay

1. The low protein concentration used in the assay is an advantage in that it keeps to a minimum the optical interference from cytochromes that occurs immediately on addition of the reagents, which thus does not affect the measurements and represents < 5% of the total absorption change with mitochondria (Boveris et al., 1972).

2. H_2O_2 formation at rates as low as 0.1 μM/min can be determined with accuracy (Chance et al., 1979) as the reaction is essentially irreversible.

3. The cytochrome-c peroxidase test offers the advantages of stability of the peroxidase-hydrogen peroxide complex and high specificity for its hydrogen donor, cytochrome c.

Limitations

It is important that samples used in this assay are free of cytochrome oxidase or cytochrome-c reductase and indeed of superoxide which also reduces cytochrome c.

3.7. Chemiluminescence and singlet oxygen

3.7.1. Introduction

Chemiluminescence has been applied to in vivo evaluation of peroxidation processes in isolated cells and organs, or intact animal organs either perfused or in situ.

The origins, significance and relevance of singlet oxygen in vivo are still unclear.

Major biological sources of singlet oxygen

(i) Systems involving interaction of lipid peroxyl radicals
The stimulation of lipid peroxidation in isolated perfused animal lung or liver is accompanied by increased light emission of low quantum yield. There is good evidence for the generation of singlet oxygen during lipid peroxidation. Its formation is attributed to the breakdown of peroxides, and hence singlet oxygen is regarded as a consequence of the lipid peroxidation rather than as an initiator of the process. This hypothesis (Sugioka and Nakano, 1976) is based on postulated reactions [3.7.1] and [3.7.2], which are by no means fully established:

$$RO_2^{\cdot} \rightarrow RO_2 - O_2R \rightarrow 2RO^{\cdot} + {}^1O_2 \qquad [3.7.1]$$
$$2RO_2^{\cdot} \rightarrow RO_2 - O_2R \rightarrow RO^{\cdot} + {}^1O_2 + RO^* \qquad [3.7.2]$$

where (RO*) is an excited alkoxyl species.

An excited group is known to be formed during the autoxidation of fatty acids, and its triplet or singlet state seems to emit in the region 420–450 nm

$$RO^* \rightarrow RO^{\cdot} + h\nu \qquad [3.7.3]$$

Light emission is proportional to the square of the concentration of lipid hydroperoxide accumulated in, for example, peroxidized mitochondrial and microsomal membranes induced by iron-mediated stress (Sugioka and Nakano, 1976) consistent with the consequence of the bimolecular reaction of lipid peroxyl radicals as a source of the chemiluminescence.

(ii) Systems involving reactions of reduced oxygen intermediates
Singlet oxygen arises in such reactions in a secondary fashion, as a product of the interaction of reductive intermediates of oxygen, such as superoxide radical, hydrogen peroxide, hydroxyl radical, as well as those systems involving an enzymatic activation of oxygen, for example, cyclooxygenase activity during prostaglandin biosynthesis.

(iii) Cellular origins of chemiluminescence
Cells in which chemiluminescence has been reported to originate include phagocytic cells, mouse spleen cells (Peterhans et al., 1980), rat thymocytes (Wrogemann et al., 1978), human NK cells (Roder et al., 1982), erythrocytes (canine) (Peerless and Stiehm, 1986), human epidermal cells (Fischer and Adams, 1985), Lettre ascites tumour cells (Mehta et al., 1985), colonic epithelial cells (Crawen et al., 1986) and Walker carcinosarcoma cells (Leroyer et al., 1987).

Reactions which generate unstable intermediates with a potential to chemiluminesce are well-known in living cells. Such examples include haem enzymes NADPH oxidase (Allen et al., 1972), cytochrome *P*-450 (Cadenas et al., 1980), cyclooxygenase (Marnett et al., 1974) and the non-haem iron-containing enzyme lipoxygenases (Veldink et al., 1977; Boveris et al., 1980).

3.7.2. *Principle of chemiluminescence*

Singlet ($'\Delta$) molecular oxygen, usually referred to as $^1O_2$, is an excited state of normal triplet oxygen ($^3O_2$) which has a long enough lifetime to be a significant reactant in biological systems (see Section 1.2). The singlet-triplet emission which occurs in the infrared (1270 nm) can be used to measure the formation of singlet oxygen, and its rate of decay. This is by far the best method for detecting singlet oxygen. A far weaker emission at 634 and 703 nm (Khan and Kasha, 1963; Seliger,

1964) can sometimes be detected, which occurs if the singlet oxygen is in high enough local concentration for reaction [3.7.4] to occur. Since this region is covered by most UV-visible spectrometers, it is still used quite extensively (Boveris et al., 1980; Cadenas and Sies, 1984).

$$^1O_2 + {}^1O_2 \rightarrow 2^3O_2 + h\nu \qquad [3.7.4]$$

3.7.3. Procedure

Detection of liberated singlet oxygen can be amplified using chemiluminogenic probes (Allen and Loose, 1976; Allen, 1981) as extrinsic substrates to which excitation energy can be transferred. These include luminol (5-amino-2,3-dihydro-1,4-phthalazinedione) and lucigenin (10,10'-dimethyl-9,9'-biacridinium dinitrate). Using these techniques, the sensitivity with which singlet oxygen generation can be detected is increased by approximately four orders of magnitude. Luminol is dissolved in spectral grade dimethyl sulphoxide and assayed spectrophotometrically: E_{mM} in water = 7.63 at 347 nm.

Sensitivity may be improved using single-photon counters as luminescence detectors.

Chemiluminescence can also be measured by reversibly modifying liquid scintillation spectrometers. Most liquid scintillation counters can easily be adapted to single photon counting but their photomultipliers are sensitive in the blue- rather than the longer wavelength spectrum of the native chemiluminescence. In most systems there is a serious technical limitation associated with native (or low-level) chemiluminescence measurement: the requirement for high cell numbers with the well-known disadvantages of the need for constant stirring or rapid nutrient depletion.

A liquid scintillation counter is actually two photon counters connected in coincidence for measuring the shower or pulse of electrons resulting from the relaxation of fluorescent molecules excited by b-particle emission. In the out-of-coincidence mode, the instrument is a single photon counter, i.e., it counts single photon events.

The luminescence intensity measurements in relative counts/minute are converted to blue (luminol) photons per minute by multiplying counts/min (cpm) by a photon conversion factor. This factor is established by calibrating the counter with an established blue-photon-emitting standard.

Systems undergoing lipid peroxidation accumulate malonyldialdehyde (see Chapters 2 and 5), as has been observed in many biological systems. The amount of malonaldehyde accumulated correlates well with the chemiluminescence intensity observed, although the chemiluminescent species and the aldehydic lipid peroxidation product are formed by different pathways and at different times during the process of lipid peroxidation.

CHAPTER 4

Transition metal complexes as sources of radicals

4.1. Introduction

We start with a general outline of the many roles played by transition metal (TM) complexes as redox agents and as sources of radicals. This is followed by a range of examples of processes thought to be of importance in biological systems. These include reactions in which radicals are produced which can lead to biological damage, as well as systems in which their formation is part of the biologically intended reaction.

4.2. Transition metal ions in biological systems

Only transition-metal ions are redox-active and hence are possible sources of radicals. Ions such as Mg^{2+}, Ca^{2+}, Zn^{2+} etc. are not important in this context. By far the most important is iron, with copper, molybdenum, cobalt and nickel also participating on a more minor scale. Of course, *any* transition metal ion may be ingested, and hence may be a fortuitous source of radical formation and damage. We focus attention on iron for illustrative purposes.

The concentrations of iron as simple solvated ions, Fe^{2+}_{aq} or Fe^{3+}_{aq}, are maintained at extremely low levels because of their damaging ability in the presence of oxygen and hydrogen peroxide. In transferrin, an important iron scavenger, the iron is well-protected and is not involved in redox chemistry. Also, in the major storage proteins, ferritin and haemosiderin, the iron is present in crystalline material inside the protein shell, and is well protected from reaction.

4.2.1. *Factors governing redox behaviour*

Redox potentials for a given metal ion are strongly controlled by the types of ligands involved, their orientation (particularly strongly controlled in proteins), the gain or loss of ligands and electron-delocalization onto ligands (Table 4.1). The rates at which ions donate or accept electrons are governed in part by these redox potentials and the concentrations of ions and substrates, but are also strongly controlled by the closest distance of approach between the reactants. Thus for two ions, M^{II} and N^{III} able to undergo electron transfer [4.1],

$$M^{II} + N^{III} \rightarrow M^{III} + N^{II} \qquad [4.1]$$

$$\boxed{M^{II}} + \boxed{N^{III}} \rightarrow \boxed{M^{III}} + \boxed{N^{II}} \qquad [4.2]$$

reaction may be very fast indeed if the ligands are small, but extremely slow if they are large [4.2] even if the reaction is equally favourable.

TABLE 4.1
Selected standard one-electron reduction potentials

Redox couple	Potential (mV) ($E_{1/2}$)
Fe^{2+}/Fe^{3+}	− 770
$Fe(II)_{EDTA}/Fe(III)_{EDTA}$	− 120
$Fe(II)_{DTPA}/Fe(III)_{DTPA}$	− 30
$Fe(II)_{citrate}/Fe(III)_{citrate}$	− 330
Cu^{2+}/Cu^{3+}	410
Cu^{+}/Cu^{2+}	− 167
Mn^{2+}/Mn^{3+}	−1510
e^{-}_{aq}/H_2O	2700
OH^{-}/OH	−1900
HO_2^{-}/HO_2	− 790
$O_2(latm)/O_2^{-}$	− 325
$O_2(lmoldm^{-3})/O_2^{-}$	− 155
HS^{-}/HS	−1150
CO_2^{-}/CO_2	2000
Cl^{-}/Cl^{0}	−2550
Cl_2^{-}/Cl_2	− 600

Sometimes the fastest mode of electron-transfer occurs when a ligand can be shared, as in M^{II}-L-N^{III}. This not only holds the partners together, but also reduces the distance of closest approach. If such bridging is involved, the reaction is said to be 'inner-sphere'; otherwise it is 'outer-sphere'. (This distinction is especially important for reactions with hydrogen peroxide (see below).) For reactions of type [4.2], if there is extensive electron-delocalization into one or more regions of the surroundings, the rates are increased because transfer can occur via the delocalizing ligands. An important example is the haem group (or porphyrins generally), which can provide a conducting pathway for the electron. We stress that there are many highly organized systems which operate via rapid electron-transfer together with proton transfer in a well-defined manner that do not involve radicals in any significant manner. However, chemical or other forms of stress may result in malfunction with resulting formation of free radicals.

4.3. Reactions with oxygen and hydrogen peroxide

Both oxygen and hydrogen peroxide are good electron acceptors. The latter has the higher affinity, but may react less rapidly because $(H_2O_2)^{\cdot-}$ is not stable, and bond-breaking probably occurs as the electron transfers, whereas for dioxygen, the

$$O_2 + e^- \rightarrow O_2^{\cdot-} \qquad [4.3]$$

$$H_2O_2 + e^- \rightarrow OH^- + OH^{\cdot} \qquad [4.4]$$

major change is a small increase in bond-length [4.3] and [4.4]. Hydrogen peroxide is a weak acid (pK_a ca. 4.6) and its conjugate base, HO_2^-, is a good electron-donor, as well as being a good nucleophile. Thus the reactions [4.5–4.10]

$$\begin{cases} M^{2+} + O_2 \rightleftharpoons M^{3+} + O_2^{\cdot-} & [4.5] \\ \ M^{2+} + O_2 \rightleftharpoons M^{3+}\text{-}O_2^{\cdot-} & [4.6] \end{cases}$$

$$\begin{cases} M^{2+} + H_2O_2 \rightleftharpoons M^{3+} + OH^- + \;^{\cdot}OH & [4.7] \\ M^{2+} + H_2O_2 \rightleftharpoons M^{3+}\text{-}OH^- + \;^{\cdot}OH & [4.8] \end{cases}$$

$$\begin{cases} M^{3+} + HO_2^- \rightleftharpoons M^{2+} + HO_2^{\cdot} & [4.9] \\ M^{3+} + HO_2^- \rightleftharpoons M^{3+}\text{-}O_2H^- & [4.10] \end{cases}$$

are all expected to occur under the correct conditions. In each case alternative steps are indicated, one involving no coordination, the other involving metal coordination. For the forms [4.5], [4.7] and [4.9], no new bonding to the metal is envisaged. This is expected for any outer-sphere mechanism where there is no available coordinations site and displacement of the ligand is difficult or impossible. However, coordination is expected to compete with single electron-transfer if there is a vacant site. The obvious example for [4.6] is oxygen coordination to haem iron. We stress that in such cases there are two extreme limiting structures, $M^{2+}\text{-}O_2$ and $M^{3+}\text{-}O_2^-$, the former being more appropriate for haemoglobin or myoglobin. The extent of internal charge transfer depends, of course, on the electron-donating power of the metal. Thus for Co(II), for example, the dioxygen complexes are often best represented as Co(III)-O_2^-.

Reaction [4.8] is expected if hydrogen peroxide interacts by an inner-sphere mechanism, as is expected; for example, for aquo complexes. In fact, reaction [4.10] may well be the final stage for reaction [4.8].

4.3.1. Fenton-type reactions

Fenton's original observation, made nearly 100 years ago, was that malic acid was rapidly oxidized by hydrogen peroxide when Fe(II) was added, but not otherwise (Fenton, 1894). Much later, Haber and Weiss (1934) suggested that such induced oxidations were due to reaction with hydroxyl radicals formed by reaction [4.7].

A large effort has been made to establish this mechanism conclusively, and many extra factors need to be considered. These include the fast reaction between Fe^{2+} and $^{\cdot}OH$ radicals and electron donation from certain R$^{\bullet}$ radicals to Fe(III). Such reactions depend on the iron complexes used, the nature of the radicals (R$^{\bullet}$) (for example, $R_2C^{\bullet}OH$ radicals are good electron donors) and relative concentrations. Clean oxidations are best achieved using aqueous solutions of the substrate and hydrogen peroxide, with slow addition of dilute Fe(II) solution to the stirred solution at room temperature.

4.3.2. Are hydroxyl radicals involved?

Although this is a most satisfying mechanism for such induced oxida-·tions, there is an important alternative, suggested originally by Bray and Gorin in 1932, involving the formation of Fe(IV) rather than ·OH. This is usually written as the formation of the aquo ferryl ion, FeO^{2+} [4.11].

$$H_2O_2 + Fe^{2+} \rightarrow H_2O + FeO^{2+} \qquad [4.11]$$

There is no reason why this should not occur, nor is the FeO^{2+} formulation any problem, although $Fe(OH)_2^{2+}$ is an obvious alternative. Although Fe(IV) complexes are rare, it is worth noting that the ferrate ion, FeO_4^{2-}, containing Fe(VI), is relatively stable, and its structure has been well-established by ESR spectroscopy (Carrington et al., 1960). Nevertheless, the evidence for ·OH participation is almost overwhelming. This includes the fact that many of the expected intermediate radicals have been detected by ESR spectroscopy, there is no kinetic salt effect for induced oxidation as would be expected if FeO^{2+} were the reactant, and products are, under correctly selected conditions, closely similar to those formed when systems known to involve attack by hydroxyl radical are compared. (In particular, radiolysis of dilute aqueous solutions can be arranged to give hydroxyl radicals as the dominating reactant.) It seems to us probable that some FeO^{2+} is formed, but that this is a minor product that is unlikely to be responsible for most of the induced oxidations.

4.3.3. Use of titanium ions

Reaction [4.12] is in many ways easier to use for induced

$$Ti^{3+} + H_2O_2 \rightarrow Ti^{3+} - OH^- + \text{·}OH \qquad [4.12]$$

oxidations than Fenton's reaction, because [4.12] is very much faster. This system has been extensively used in flow ESR studies of a very wide range of unstable radicals. First developed by Dixon and Norman (1962) the method has been greatly extended by Gilbert, Norman and their co-workers, and more recently by Gilbert and Jeff (1988). Again the evidence is overwhelming, that these reactions involve the intermediate formation of ·OH radicals.

4.3.4. Fenton-type reactions in lipids

As stressed throughout this book, lipid oxidation is one of the more important biological radical processes. Being a chain reaction, only low concentrations of initiators are required, and it has been argued that an important initiator may be a Fenton type reaction.

The main problem with this concept is one of solubility. The site of attack is the double-bonded part of the lipids, embedded in the centre of the membrane. Not only are Fe(II) complexes unlikely to be present, but also H_2O_2 is unlikely to be present in the inner region of the phospholipid double layer. Thus, the probability of there being sufficient of each to initiate autoxidation seems to us to be very low indeed. Nevertheless, catalysis by metal ions does occur. Possibly, membrane proteins are involved in some way.

4.3.5. Can copper (I) ions act as a source of hydroxyl radicals?

Since Cu(I) ions are usually powerful electron donors, the reaction with H_2O_2 to give ·OH radicals is highly probable. However, the reaction [4.13] is less useful as a method of inducing oxidation because the disproportionation reaction to give Cu(II) and copper metal ensures that [Cu(I)] is always low.

A strong case against this reaction [4.13] has been given by Johnson and co-workers (1988), who favour Cu^{3+} as an intermediate. This is an exactly analogous problem to that outlined above for

$$Cu^+ + H_2O_2 \rightarrow Cu^{2+} - OH^- + \cdot OH \qquad [4.13]$$

reaction with ferrous complexes. Clearly both possibilities exist, and at least in some circumstances, involvement of $\cdot OH$ radicals seems to be most probable. We do not feel that the evidence against $\cdot OH$ radical involvement is clearly established. Certainly, the radicals formed are those expected for $\cdot OH$ radical attack and $Cu^+{}_{aq}$ can hardly avoid participating in reaction [4.13]. However, ligands which greatly stabilize the Cu(I) state may also inhibit such a reaction. The recent work of Gilbert and Stell, in which they show in a compelling way that Cu(I) reacts with monopersulphuric acid to give $\cdot OH$ radicals is especially significant to this discussion. In these reactions there is no need to invoke Cu^{3+} formation (Gilbert and Stell, 1990).

4.4. Transition metal ions and oxygen radicals in biological systems

The reasoning given above applies equally to the biological field, and the same types of ambiguities arise. Rather than attempting further generalizations, we prefer to describe certain case histories. A separate section is devoted to the ferryl-haem systems since these are well-established as important intermediates in certain reactions of hydrogen peroxide [4.5]. Finally, in Section 4.6 various experimental techniques for studying iron and copper in biological systems are outlined.

4.4.1. Generation of superoxide ions by xanthine – xanthine oxidase

The conversion of O_2 into $O_2{}^{\cdot-}$ with release of this ion into the medium is now well-established. The most direct proof is the elegant work of Bray et al. (1977) using their rapid-freeze technique coupled with

ESR spectroscopy. Although $O_2^{\cdot-}$ has never been detected by ESR spectroscopy in fluid aqueous solutions, it can be detected in frozen systems. Very rapid freezing can be achieved during or after mixing using a fine spray impinging on a cold metal plate. Spraying into liquid nitrogen at its boiling point results in much slower cooling, and is not effective. Their results clearly establish $O_2^{\cdot-}$ formation, and the system xanthine–xanthine oxidase is used quite widely as a source of superoxide ions. It must be borne in mind, however, that hydrogen peroxide is also released under these conditions, which may complicate the results and interpretation.

4.4.2. Generation of hydroxyl radicals by ferredoxin systems

Our second example, which seems to provide unambiguous evidence for $\cdot$OH radical involvement, is taken from one of the many interesting ESR studies of Ron Mason and his co-workers (Morehouse and Mason, 1988). They used the spin trap DMPO in an aerobic incubation of ferredoxin + ferredoxin:NADP$^+$ oxidoreductase + NADPH. The major ESR signal was from the DMPO-OH adduct with a minor contribution from the superoxide adduct. As stressed above, care has to be taken that DMPO was not directly involved in a redox reaction to give the $-$OH adduct. This was nicely established by using $^{17}O_2$. Normal oxygen contributes no hyperfine splitting (^{16}O is non-magnetic), but ^{17}O contributes a six-line splitting ($I = 5/2$) which was clearly seen. Hence the $\cdot$OH adduct was derived from $^{17}O_2$ and not from the addition of water to DMPO$^{\cdot+}$. The results were interpreted in terms of reaction [4.14].

$$O_2 \xrightarrow{e^-} O_2^{\cdot-} \xrightarrow[H^+]{e^-} H_2O_2 \xrightarrow{e^-} {\cdot}OH + OH^- \qquad [4.14]$$

All three ingredients were necessary for radical formation. Hydrogen peroxide greatly enhanced the yield, whilst catalase inhibited the $\cdot$OH adduct formation. The system was able to induce the oxidation of ethanol, and reaction rates agreed with those expected for $\cdot$OH radical attack.

In this reaction, ferredoxin is reduced by ferredoxin:NADP$^+$ oxidoreductase, and this reduces oxygen to hydrogen peroxide. However, the results showed that the reduced ferredoxin does not react itself with H_2O_2 to give ·OH. It is suggested that iron impurities, reduced to the ferrous form, are responsible for a Fenton-type reaction with H_2O_2. This was supported by the addition of chelators such as DTPA, which greatly enhance the rate of ·OH radical formation. It is stressed that ·OH radicals are formed in good yields even from extremely low Fe(II) concentrations (ca. 0.03 μM). Strong arguments are given in favour of ·OH radical involvement. Again, ferryl derivatives may be formed but they do not appear to be either a major product or to be the source of the ·OH adducts or the induced oxidations.

The results are especially important in showing that although the enzyme system itself does not produce ·OH radicals, it *does* produce H_2O_2 and this will seek out Fe(II) complexes even in very low concentrations.

4.5. Iron determination

Studies of reactions such as those discussed above rest upon analytical techniques for the determination of metal ions. Some specific techniques are given below for the measurement of total iron levels, non-haem iron and for haem iron.

4.5.1. Total iron determination by atomic absorption spectroscopy

Total iron levels (free + bound) in a range of biological systems such as urine, plasma, membranes, solution etc. can be determined applying atomic absorption spectroscopy.

Reagents
Concentrated nitric acid – 70% spectrosil (BDH special for atomic absorption spectroscopy – iron level < 0.2 ppm).

Procedure
(i) Liquid samples. 1 ml samples, in buffer, of known protein concentration are taken in triplicate into large diameter pyrex boiling tubes. In the case of urine samples, known undiluted volumes are taken, usually of 1 ml.

The samples are heated in a dry block at 170°C until the sample is completely dry. A 2 ml aliquot of fuming nitric acid is carefully added using a glass calibrated pipette, the tubes covered with a large glass marble to prevent evaporation and the sample is heated at 170°C to digest the sample. After 20 min the marble is removed, and the heating continued until the nitric acid has completely evaporated.

The digested material is resuspended in a further 2 ml aliquot of 70% nitric acid and utilized for atomic absorption spectroscopy. Blanks containing the same volume of the appropriate buffer are similarly taken through the entire process.

Iron levels are measured at 248.3 nm in the atomic absorption spectrometer.

(ii) Solid samples. For measurement of faecal iron excretion, for example, a known weight of sample is digested with twice its weight of water. Faecal homogenates are incubated with 5 ml of concentrated nitric acid for every 10 g. Samples are allowed to digest for 20 min at ca. 95°C, avoiding evaporation as before. After cooling, incubations are filtered and the filtrate is transferred to a volumetric flask and diluted to 100 ml.

4.5.2. Chemical methods for non-haem iron

4.5.2.1. Bleomycin assay
Bleomycin, an antitumour antibiotic, binds iron in the reduced, divalent state to its secondary amide group. It binds 'low molecular weight' iron ions (often loosely referred to as 'free' iron) in forms that catalyse free radical reactions (Gutteridge et al., 1981) but *not* iron bound within native, functional protein structures, such as ferritin, transferrin, lactoferrin, haemoglobin, myoglobin, cytochromes etc. The resulting bleomycin-iron complex is capable of degrading DNA

in the presence of the reducing agent, ascorbic acid. When the three components, bleomycin, ascorbate and DNA, are present in excess, the amount of DNA degradation can be used as a marker for the concentration of available or 'free' iron ions as described below. The bleomycin assay has been applied to assess the levels of available, nontransferrin-bound iron in plasma samples in patients with iron overload disease (Peters et al., 1985), with acute leukaemia before and after drug chemotherapy (Halliwell et al., 1988), as well as patients with rheumatoid arthritis (Winyard et al., 1987).

Principle of the assay
The assay is based on the property of bleomycins to act as anti-tumour antibiotics by binding to DNA, especially adjacent to guanosine residues. They cause single-strand and some double-strand breakages and degradation of the deoxyribose sugars, leading to products that react with thiobarbituric acid, forming a pink complex (Fig. 4.1). (λ_{max} 532 nm.) The concentration of the pink complex is dependent on the extent of destruction of the deoxyribose substituents of the DNA, which, in turn, is dependent on the amount of available iron present in the system.

Treatment of DNA with iron chelators such as EDTA, DETAPAC or desferrioxamine protects it from degradation. Thus, the destruction of DNA by bleomycin is iron-dependent. The breakdown of bleomycin may be initiated by a ferryl radical species or a hydroxyl

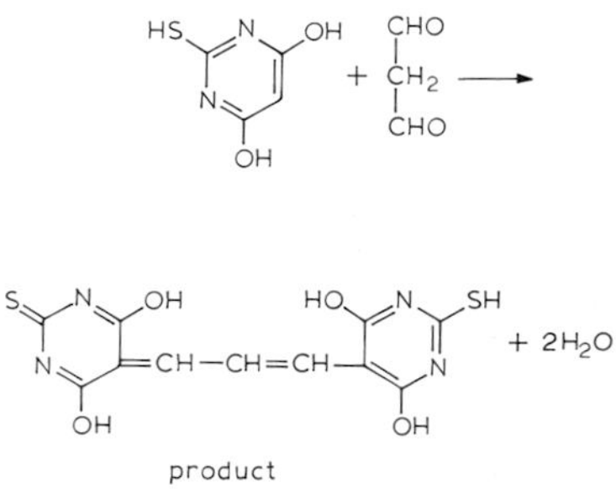

Fig. 4.1. Interaction between thiobarbituric acid and malonaldehyde.

radical. A range of free radical scavengers do not protect DNA against bleomycin induced damage. This includes superoxide dismutase, catalase, caeruloplasmin, thiourea, mannitol, dimethyl sulphoxide.

Presumably, if the reaction is mediated by the formation of a hydroxyl radical, it is formed so close to the DNA that it cannot be relatively more accessible to the added antioxidant to be scavenged effectively.

Preparation of reagents

1. All glassware must be rinsed with iron-free distilled water before use.

2. All reagents must be prepared in iron-free distilled water.

3. Reagent A:
   DNA (calf thymus) is dissolved in distilled water at a concentration of 1 mg/ml by gentle mixing and allowed to stand for at least 12 h before use.

4. Reagent B:
   Bleomycin sulphate (Sigma) is dissolved in distilled water at a concentration of 0.6 mM (1.5 units/ml). (A mask should be worn when opening the sealed ampoules of bleomycin to protect against inhalation of the dust.)

5. Reagent C:
   Solutions of ascorbic acid (analar) must always be freshly prepared. 0.7 g are dissolved in 10 ml distilled water, shaken with 0.4 g Chelex resin (to ensure iron removal) and centrifuged at 2000 × *g* to remove the resin. It is essential that *all* the chelex is removed at this centrifugation stage. Aliquots of the supernatant (1 ml) are diluted 50-fold with water.

6. Reagent D:
   Thiobarbituric acid (TBA) is dissolved in 0.05 M sodium hydroxide at a concentration of 1% (w/v).

7. 50 mM magnesium chloride solution/10 mM hydrochloric acid/ 25% (w/v) hydrochloric acid/0.1 M EDTA.

Procedure
The reagents should be added to a *new* plastic tube, in the sequence defined in Fig. 4.2 and the *procedure, as depicted, followed precisely.*
    After the addition of reagent 11 on completion of procedure in vitro, the solution is transferred to acid-washed glass tubes, heated at

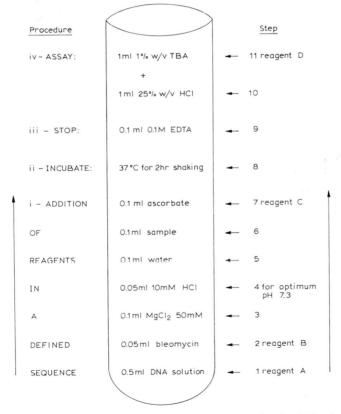

Procedure		Step
iv – ASSAY:	1ml 1% w/v TBA	11 reagent D
	+	
	1ml 25% w/v HCl	10
iii – STOP:	0.1 ml 0.1M EDTA	9
ii – INCUBATE:	37 °C for 2hr shaking	8
i – ADDITION	0.1 ml ascorbate	7 reagent C
OF	0.1ml sample	6
REAGENTS	0.1ml water	5
IN	0.05ml 10mM HCl	4 for optimum pH 7.3
A	0.1ml MgCl₂ 50mM	3
DEFINED	0.05ml bleomycin	2 reagent B
SEQUENCE	0.5ml DNA solution	1 reagent A

Fig. 4.2. Procedure for the bleomycin assay for non-haem iron. Reagents must be added strictly in the order of the direction of the arrow.

100°C for 10 min, cooled and centrifuged at 2000 × g for 10 min. The supernatant is carefully removed, and the absorbance measured at 532 nm.

Controls of two types are set up simultaneously, consisting of an identical procedure to that described above for the samples except that one of the controls will contain all components *excluding the sample*, the other will contain all components *excluding the bleomycin* but including the sample.

A standard curve using a range of concentrations of iron(III) chloride in iron-free distilled water must be obtained and used to quantify the amount of iron involved in the reaction.

At a 10 μM concentration the absorbance measured in the assay will be approximately 0.5 at 532 nm. The assay is reported to be linear up to 50 μM iron salt.

Precautions:

Bleomycin sulphate must be handled extremely carefully, as inhalation of the powder has been suggested to cause lung damage.

Although this appears to be a straightforward assay, it is in fact extremely difficult to obtain reliable, reproducible results. In the authors' opinion it is an assay to be avoided unless no other is available.

In order to facilitate the assay the following precautions are essential:

The reagents must be added in the order defined in the procedure above.

This is a highly sensitive assay, hence all traces of contaminating iron must be removed from all the reagents. Any iron contamination will show up as a high blank in the assay. Where chelex resin is applied for iron removal, it is essential to ensure that all chelex must be subsequently removed to prevent contamination of the assay.

4.5.2.2. Ferrozine assay

Non-haem iron in solution, plasma or associated with membranes, may be assessed utilizing the ferrozine assay (Ceriotti and Ceriotti, 1980).

Principle

Ferrozine {3-(2-pyridyl)-5,6-bis(4-phenylsulphonic acid)-1,2,4-triazine} (Fig. 4.3) reacts with iron in the iron(II) ferrous state to form an intense pink complex which can be measured spectrophotometrically (λ_{max} = 562 nm). The assay can be performed at pH 1.6 to eliminate interference from turbidity at pH values close to the isoelectric points of proteins which may be present in biological samples under investigation, e.g., serum, and in some instances to decrease interference in the assay from copper (see below). In determinations in which such problems are not considerations, the assay may be carried out at pH 7.4. (This might be more relevant for the assessment of membrane-bound iron under in vivo conditions.)

pH 1.6 assay

REAGENTS

1% (w/v) ascorbic acid (prepared immediately prior to use) and 0.25% thiourea in 0.1 M HCl. Both these reagents are dissolved in water and HCl subsequently added to 0.1 M concentration.

 Glycine buffer: 0.2 M glycine in 5 mM HCl, pH 4.15. 20 mg/ml ferrozine in water (prepared freshly on the day). Standard iron stock solution for calibration consisting of 0.45 mM iron chloride, i.e., 2.23 mM iron(II) in glycine buffer (prepared immediately prior to use, since iron(II) salts oxidize very readily in air).

PROCEDURE FOR ASSAY AT pH 1.6

1. Sample assay.

2 ml ascorbate/thiourea solution are mixed in a glass tube (or directly

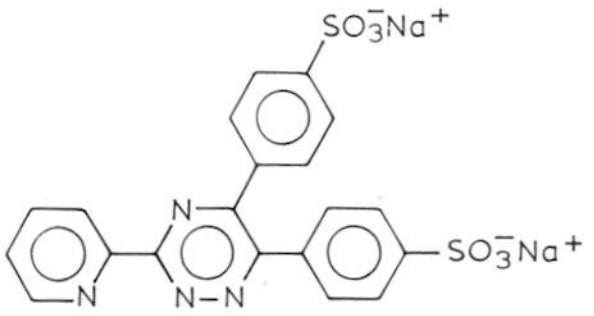

Fig. 4.3. Structure of ferrozine: 3-(2-Pyridyl)-5,6-bis(4-phenylsulphonic acid)-1,2,4-triazine, sodium salt.

into cuvettes) with 0.5 ml of 0.2 mM glycine buffer and 0.1 ml ferrozine solution; a 0.2 ml aliquot of the sample under investigation is added. After 20 min an aliquot is transferred to a glass or quartz cuvette and the absorbance is measured at 562 nm over a time-scale of 0–3 h, and at 24 h and 48 h. For membrane systems in order to ensure access of reagents to membrane-bound iron it is preferable to solubilize membranes using a detergent (Hartley et al., 1990); 0.2 ml of 10% (w/v) sodium dodecyl sulphate is incorporated into the sample assay mixture.

2. Blank determination.

At the same time as the test solution two blank determinations are carried out, the first consisting of the assay system as above but in the absence of ferrozine with 0.1 ml water in lieu. This allows for the contribution to the absorbance from haem proteins or other absorbing components which may be present; usually the absorbance of this blank is of the order of 0.005. The second consists of the assay system in the absence of the sample, replaced by 0.2 ml water, to allow for interaction of ferrozine with reactable iron in the reagents which evaded removal prior to assay.

The average of the blank 0–3 h readings is subtracted from all the sample readings.

3. Standard curve.

In order to determine the relationship between the absorbance at 562 nm and the iron content of the iron(II)-ferrozine complex a standard curve is prepared by taking 0.2 ml aliquots of the standard iron solutions, 0–125 mg/l, containing amounts ranging from 0–25 μg iron(II).

SUMMARY OF PROCEDURES

(i) at *pH 1.6*
Combine reagents in a sequence shown:
1. 2.0 ml of ascorbate/thiourea solution
       (1% (w/v), 0.25 % (w/v), respectively, in 0.1 M HCl).
2. 0.2 ml of SDS solution (10%, w/v).

3. 0.5 ml of 0.2 M glycine buffer, pH 4.15.
4. 0.1 ml of 20 mg/ml ferrozine solution (or water for non-ferrozine blank).
5. 0.2 ml sample or standard or water (non-sample blank).
6. Mix all reagents in a glass tube (or directly into a cuvette).
7. Transfer to cuvette and measure the absorbance at 562 nm.

(ii) at *pH 7.4*
1. 2.1 ml of 5 mM phosphate-buffered saline, pH 7.4.
2. 0.2 ml of SDS solution.
3. 0.4 ml of ascorbate.
4. 0.1 ml of ferrozine solution (or water for non-ferrozine blank).
5. 0.2 ml of sample or standard or water (for non-sample blank).

Sensitivity
The sensitivity of the assay can be judged from utilizing an iron concentration of 3.3 μM or 3.7 μg iron in 0.2 ml, giving an absorbance of 0.08 at 562 nm.

Results
The test is illustrated with results taken from experiments on membranes of erythrocytes from patients with sickle cell anaemia, which tend to retain increased levels of membrane-associated iron, compared with those of normal erythrocytes. Some results are shown in Table 4.2.

Sources of interference
The major source of interference and inaccuracy is iron contamination. It is thus essential to ensure that all the reagents, especially the ascorbate and the hydrochloric acid, are iron-free. Copper ions may also interfere in this assay (Duffy and Gandin, 1977). This can be eliminated by incorporating thiourea which, at a low pH, forms a stable colourless complex with copper without affecting the reaction of ferrozine with iron, thereby reducing the colour formation of ferrozine with copper. The affinity of thiourea for copper is maximal in the low pH range used here.

TABLE 4.2.

Total, haem and non-haem iron levels in erythrocytes

	Membrane-bound iron retention (nmol/mg protein)		
	Haem iron[a]	Non-haem[b]	Total iron[c]
Normal	1.69 ± 0.73	0	1 ± 0.6
Median (range)	1.67 (0.53–2.56)	0	1 (1–2)
	(6)	(6)	(13)
Sickle	1.58 ± 0.61	22 ± 20	25 ± 23
Median (range)	1.57 (0.35–2.58)	16 (5–78)	15 (1–93)
	(13)	(27)	(37)

[a]Method of Morrison, 1965.
[b]Method of Ceriotti and Ceriotti. The data indicate the large variation in the extent of non-haem iron retention in the erythrocyte membranes of patients with sickle cell anaemia.
[c]Atomic absorption spectroscopy.

Figures in brackets represent the numbers of patients/controls ± standard deviation.

In the absence of thiourea, interference by copper is significant in plasma when testing serum, producing very significant positive errors especially at mean and low iron and copper concentrations. This combination of conditions is more common in such diseases as malignant lymphoma and leukaemia.

4.5.3. Chemical methods for haem iron

The levels of haem proteins in biopsy material, in plasma and bound to membranes can be determined to a level of 0.01 μg, applying the following spectrofluorimetric method (Morrison, 1965).

Principle

The method is dependent on the conversion of the haem moiety to its fluorescent porphyrin derivative by incubation with oxalic acid.

REAGENTS

Oxalic acid 31.51 g in 250 ml water.

Quinine sulphate 0.1 μg/ml in 0.1 M sulphuric acid.

Standard haem solution prepared from haemoglobin solution 5 μM final concentration; either human haemoglobin (Sigma) or red cell lysate (after removing membranes by centrifugation) can be used. The concentration of the standard haemoglobin solution is standardized using the cyanmethaemoglobin method.

Procedure for the determination of membrane-bound haem

1. Membranes to be assessed for membrane-bound haem iron are diluted in 5 mM phosphate buffer, pH 7.4, to give a protein concentration of 0.3 – 1.2 mg/ml.

2. 4 ml oxalic acid solution is added to 100 μl of diluted sample at 110°C for 90 min and then allow the tubes to cool.

   Set up standards in the same way taking 0 – 0.5 nmol haem in 100 μl volume.

3. The fluorimeter is calibrated using the quinine sulphate solution 0.1 μg/ml, excitation wavelength 360 nm and setting the emission intensity to 86 at 450 nm.

4. The sample is scanned in the fluorimeter from 560 – 680 nm on excitation at 410 nm. The peak height is measured at 603 nm. Typical results from normal human erythrocyte membranes and sickle erythrocyte membranes compared with non-haem iron levels and total iron measurements are given in Table 4.2 (Hartley and Rice-Evans, 1989).

4.6. Spectroscopic identification of ferryl haem proteins

Introduction

Various haem proteins such as horseradish peroxidase, cytochrome-*c* oxidase and cytochrome peroxidase have been shown to be oxidizable to states which correspond formally to Fe(IV) (Compound II) and Fe(V) (Compound I) oxidation levels (Yamada and Yamazaki, 1974;

Witt and Chan, 1987; Rao et al., 1988) described as 'ferryl' states. The Fe(IV) state has been designated as an iron-oxo complex (FeO^{2+}); the Fe(V) state is a similar complex, the extra electron being lost from the porphyrin π-system ($P^{\cdot+}$-FeO^{2+}) (Felton et al., 1976; Rakshit et al., 1976).

Metmyoglobin also reacts with hydrogen peroxide to give relatively stable species that have ferryl-like properties, ferryl myoglobin, the haem iron being one oxidizing equivalent above that of the Fe(III) in metmyoglobin (George and Irvine, 1952; Yonetani and Schleyer, 1967; Petersen et al., 1989) and confirmed by Mossbauer spectroscopy (Maeda et al., 1973); a transient free radical form of ferryl myoglobin, designated $^{\cdot}$Mb(IV), is also formed (King et al., 1967).

$$HX \sim Fe(III) + H_2O_2 \rightarrow {}^{\cdot}X\text{-}[Fe(IV)\text{=}O] + H_2O$$

ESR studies have confirmed that when metmyoglobin (and oxymyoglobin) react with hydrogen peroxide, ferryl myoglobin is the major product (Petersen et al., 1989) but the proportion depends on the relative concentrations of the haem protein to hydrogen peroxide (Whitburn 1987, 1988). Despite some ambiguity regarding the molecular mechanism for the formation of Mb(IV), its formulation as [HX-Fe(IV)-OH] or [HX-Fe(IV)=O], as well as the designation of the ESR-detectable transient free radical form of the protein as the porphyrin radical cation [$^{\cdot}$X-Fe(IV)-OH], has been substantiated by a variety of studies. The location of the radical in the globin rather than on the porphyrin ring renders the radical accessible for the initiation of oxidative damage to a variety of susceptible components; in addition, the structural preservation of myoglobin itself against radical damage in vivo may be attributed to the existence of a means of transferring free radical sites from the vicinity of the haem to a tyrosine residue which is accessible to external reducing agents.

It has been proposed that the second oxidizing equivalent is accepted by tyrosine-103 on the surface of the protein (Ortiz de Montellano, 1983) forming a tyrosine phenoxyl radical which rapidly takes up oxygen forming the peroxyl species (Davies, 1990) as depicted in Fig. 4.4.

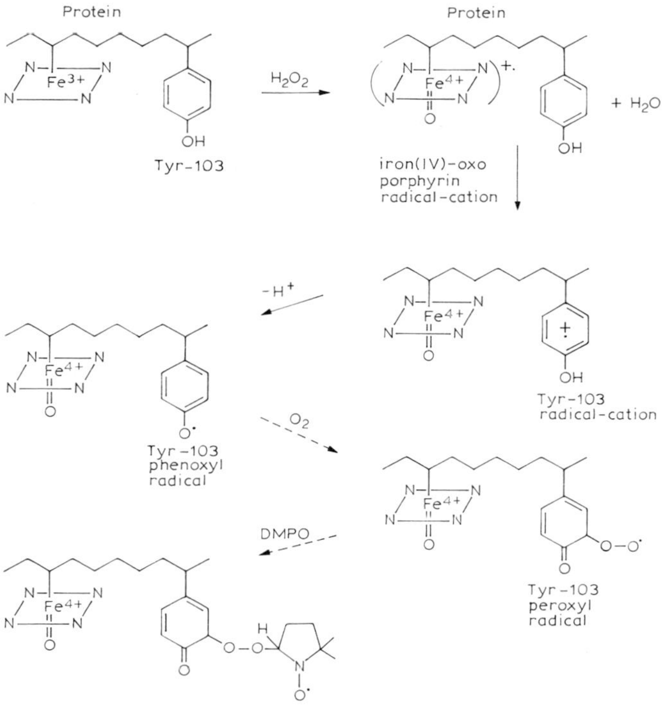

Fig. 4.4. Scheme for postulated formation of the tyrosine phenoxyl and peroxyl radical in ferryl myoglobin.

Identification of ferryl myoglobin species

It is important to bear in mind that ferrylmyoglobin has two forms: the radical species as described above, and the non-radical iron-oxo ferryl form in the iron(IV) state.

The former can be detected by ESR spectroscopy (Fig. 4.5), by spin-trapping with DMPO or by stopped-flow analysis (Chapter 1).

Ferryl myoglobin can be identified spectrophotometrically, although it is not possible to differentiate between the radical and non-radical state using this technique. Whereas metmyoglobin shows characteristic peaks at 535, 585 and 630 nm, ferrylmyoglobin has peaks

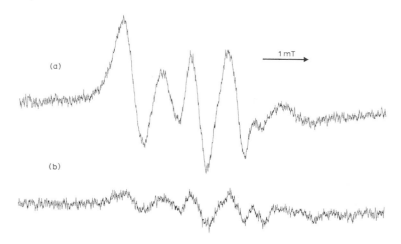

Fig. 4.5. ESR spectrum of the ferryl myoglobin radical (Turner et al., 1990). ESR spectrum observed on reaction of **(a)** metmyoglobin (225 μM) and **(b)** oxymyoglobin with 250 μM hydrogen peroxide in the presence of 25 mM DMPO at pH 7.4 under normoxic conditions.

at 540, and 580 nm, with isosbestic points at 520 and 623 nm. The two spectra differ most at 556 nm (Fig. 4.6).

The total haem concentration and the relative proportions of ferryl-met- and oxy- can be assessed applying the Whitburn algorithm (Whitburn, 1987):

$$[\text{oxyMb}] = 2.8\,A_{490} - 127\,A_{560} + 153\,A_{580}$$
$$[\text{metMb}] = 146\,A_{490} - 108\,A_{560} + 2.1\,A_{580}$$
$$[\text{ferrylMb}] = -62\,A_{490} + 242\,A_{560} - 123\,A_{580}$$

These equations are applicable to ferryl myoglobin radical production in chemical systems, but the results are less clear when applied to cellular systems such as cardiac myocytes. A better spectroscopic approach here is to observe the shifts in the Soret region to detect qualitatively the presence of ferryl myoglobin (Turner et al., 1991). Fig. 4.6 shows the comparison between the visible spectra of metmyoglobin

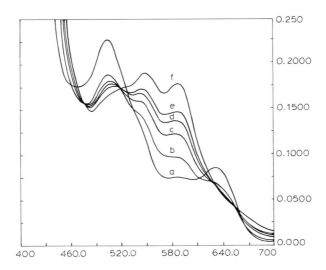

Fig. 4.6. Spectroscopic identification of ferryl myoglobin. The activation of metmyoglobin to ferryl myoglobin by hydrogen peroxide (× 1.25 molar excess): the development of the characteristic peaks of ferryl myoglobin as a function of time (minutes): **(a)** 0, **(b)** 0.5, **(c)** 2.5, **(d)** 4.5, **(e)** 6.5 and **(f)** 8.5. The development of the characteristic peaks for ferryl myoglobin around 550 and 580 nm is shown, with a decrease in the shoulder at 630 nm, characteristic of metmyoglobin, and the peak at 515 nm.

and ferryl myoglobin with the proportions of each component as calculated using the Whitburn equations (Fig. 4.7).

Preparation of pure met- and oxymyoglobin
If chemical studies are set up to investigate the formation and reactions of ferryl myoglobin, the metmyoglobin and oxymyoglobin must be purified.

Horse heart myoglobin (ferric form, salt free, lyophilized) is purified by oxidation followed by separation, using column chromatography. A 1 mM solution of metmyoglobin is prepared and treated with an equimolar solution of ferricyanide.

After 15 min at room temperature, the solution is carefully layered onto a Sephadex G-15 column, pre-equilibrated with phosphate buffer, pH 7.4, and the eluted solution is collected.

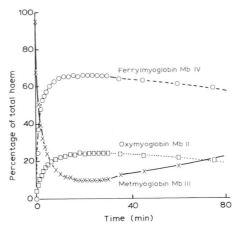

Fig. 4.7. Formation of ferryl, met- and oxymyoglobin on activation of metmyoglobin (20 μM) by hydrogen peroxide ($\times$ 1.25 molar excess). Proportion calculated applying the Whitburn algorithm ($n=4$; SD $<3\%$).

To prepare the oxygenated divalent form of myoglobin, a 1 mM solution of metmyoglobin in phosphate buffer, pH 7.4, is treated with a 10-fold excess of sodium dithionite. The resulting solution is dialysed against 100 vol. of distilled water for 2 h at 4°C and are immediately stored-frozen.

CHAPTER 5

Investigation of the consequences of free radical attack on lipids*

5.1. Initiation of lipid peroxidation and conjugated diene formation

5.1.1. Introduction

Lipid peroxidation is a free radical-mediated, chain reaction resulting in the oxidative deterioration of polyunsaturated fatty acids (PUFAs) defined for this purpose as fatty acids that contain more than two double covalent carbon-carbon bonds. Singlet oxygen can produce lipid hydroperoxides in unsaturated lipids by non-radical processes (Pryor and Castle, 1984), but the reaction usually requires a radical mechanism (Porter, 1984). Polyunsaturated fatty acids are particularly susceptible to attack by free radicals. Lipid peroxidation is a complex process, and three distinct phases are recognized: (a) initiation, (b) propagation and (c) termination (see Fig. 2.10).

Initiation of lipid peroxidation can proceed by hydrogen abstraction (Pryor and Castle, 1984) or possibly by an addition reaction (Willson, 1979; Porter, 1984). Naturally occurring PUFAs usually contain methylene interrupted structures, e.g., R-CH=CH-CH$_2$-CH=CH-R$_2$. The presence of carbon-carbon double bonds weakens carbon-hydrogen bonds in the adjacent methylene group and facilitates hydrogen atom abstraction thus:

*The authors are grateful to Dr. P.T. McCarthy for his major contribution to this chapter.

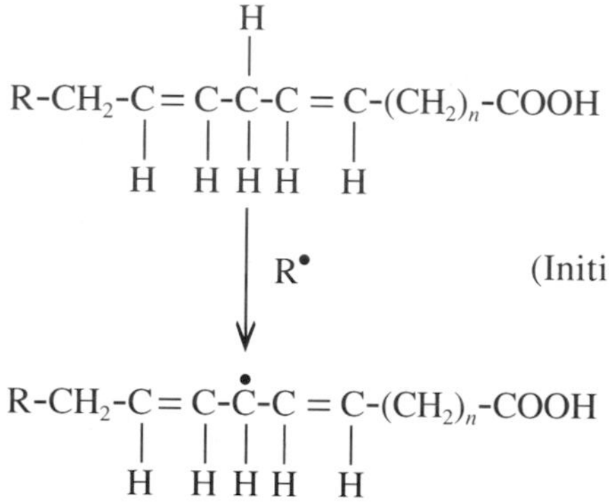

(Initiation by H-atom abstraction)

The carbon-centred radical formed is stabilized by spontaneous molecular rearrangement to form a diene-conjugate (Farmer et al., 1941; Farmer and Sundralingham, 1942):

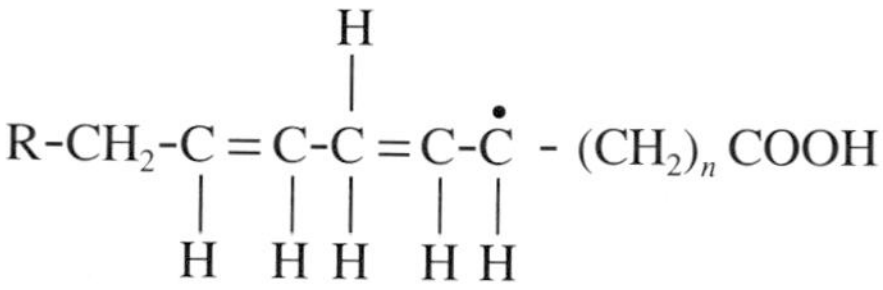

The life-span of this diene-conjugated radical may be very short, but the diene configuration is retained in more stable products including hydroperoxides, alcohols derived from the reduction of lipid hydroperoxides, and fatty acids. The spectra of molecules expressing the diene configuration are characterized by an intense light absorption (the K band) in the region of 215–250 nm (Gillam et al., 1931; Edisbury et al., 1933). Peroxidized fatty acids absorb strongly at 233 nm and exhibit a secondary absorption maximum due to ketone dienes in the region of 260–280 nm (Fig. 5.1), which also demonstrates that this method can be used as a measure of the peroxidation of tissue lipids that are exposed to peroxidative conditions in vitro. Thus the measurement of conjugated dienes provides a fast, accurate and convenient method for the assessment of lipid peroxidation. Conjugated dienes may be measured by ultraviolet spectroscopy, or, if a more pre-

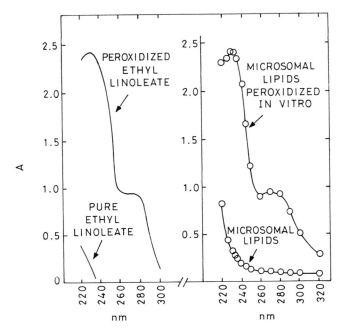

Fig. 5.1. Conjugated diene ultraviolet absorption of peroxidized ethyl linoleate and rat liver microsomal lipids peroxidized in vitro. (Recknagel and Glende, 1984.)

cise and selective measurement of specific dienes is required, by high-performance liquid chromatography.

5.1.2. Detection of conjugated dienes by ultraviolet spectroscopy

The method given for the detection of conjugated dienes in lipids from body tissues and subcellular organelles is derived from Recknagel and Glende (1984). The major disadvantage of the technique is that the absorption maximum of the dienes is superimposed on the end absorption of non-conjugated, non-peroxidized lipids which also absorb in the ultraviolet spectral region. When the proportion of endogenous lipids that are actually peroxidized is low, the end absorption of the un-peroxidized lipids masks the absorption of the dienes. However,

errors in measurement can be lessened by utilizing difference spectroscopy using un-peroxidized lipids as the spectrophotometric blank, or
by using double-beam UV spectrophotometers.

5.1.2.1. Method

REAGENTS:

1. Chloroform/methanol, 2:1 (v/v) prepared from solvents of the
   highest purity, i.e., Analar, Aristar or HPLC grades.
2. Cyclohexane: spectrophotometric grade.
3. Oxygen-free nitrogen: (prepared by flushing 'commercial gas' over
   hot, reduced copper mesh or by passage through an oxygen trap).

DETAILS OF PROCEDURE:

(a) Isolated tissues or organelles.

1. Isolated tissues (2–3 g) (or organelles from 2–3 g of tissue) are
   transferred with 6–7 ml methanol to a graduated 40 ml stoppered
   centrifuge tube.
2. The volume is adjusted to 7 ml with methanol, 14 ml of chloroform
   added, and mixed thoroughly.
3. The mixture is allowed to stand at room temperature for 10 min,
   after which it is centrifuged at $260 \times g$ for 10 min.
4. The supernatant fluid is decanted into a similar centrifuge tube, the
   volume adjusted to 30 ml by addition of chloroform/methanol
   (2:1,v/v) and 10 ml of distilled water is then added, followed by
   gentle mixing by inversion.
5. The contents are again centrifuged for 10 min and the upper meth-
   anol/water phase and any fluffy material at the interface aspirated
   to waste.
6. 2 ml of the remaining chloroform layer is transferred to a clean
   tube, the solvent completely evaporated under nitrogen, and the
   residue re-dissolved in spectrophotometric grade cyclohexane.
7. For measurement of conjugated dienes, the solution is scanned between 220 and 300 nm against a control or cyclohexane blank (Fig.
   5.1). Absorbance measurements are made at a uniform base of 1
   mg of lipid/ml of cyclohexane; for this purpose total lipids may be

determined by the method of Chiang et al. (1957). The process of tissue isolation, lipid extraction and diene measurement should be performed without an overnight delay to avoid auto-oxidation in vitro.

5.1.3. Detection of conjugated dienes by high-performance liquid chromatography

Qualitative and quantitative information about the main diene-conjugated compounds formed during lipid peroxidation can be obtained using HPLC techniques (Cawood et al., 1983; Iversen et al., 1984, Iversen et al., 1985; Harrison et al., 1985). These workers suggested that 95% of the 'total' diene conjugation in fresh human and animal serum, tissues and fluids can be accounted for by a single fatty acid, octadeca, 9-cis, 11-trans-dienoic acid [18:2(9-cis, 11-trans)] which contains no other oxygen than the carboxyl function. The generation of 18:2(9-cis, 11-trans) from 18:2(9,12-cis) linoleic acid by a free radical mediated mechanism requires the presence of protein and is thought to involve an interaction between a carbon-centred radical with protein thiol residues, thus (Wickens and Dormandy, 1988):

$$18:2(9,12\text{-}cis) \xrightarrow{\text{hydrogen abstraction}} 18:2(9,12\text{-}cis)$$
$$18:2(9,12\text{-}cis) \xrightarrow{\text{diene conjugation}} 18:2(9,11\text{-}cis)$$
$$18:2(9,11\text{-}cis) + \text{Protein-SH} \rightarrow 18:2(9\text{-}cis,11\text{-}trans) + \text{Protein-S}^{\bullet}$$
$$2\,\text{Protein-S}^{\bullet} \xrightarrow{\hspace{4cm}} \text{Protein-S-S-Protein}$$

but other mechanisms may apply.

Several clinical studies have provided circumstantial evidence that a free-radical mechanism might be responsible for increased levels of 18:2(9-cis,11-trans) in blood. Significantly raised levels of the isomer have been detected in the serum of chronic alcoholics (Fink et al., 1985; Szebeni et al., 1986). It is suggested that a free radical-dependent detoxifying mechanism is invoked by heavy alcohol loads in such persons and similar high levels have also been found in the plasma of persons poisoned with paraquat, the toxicity of which is known to involve free radicals (Crump et al., 1988), in chronic biliary cirrhosis

(Fink et al., 1985) and in pre-eclamptic toxaemia (Erskine et al., 1985). Initial studies of human biopsy samples, of exfoliated cells from normal cervix, and of those with colposcopic and cytological evidence of precancer suggested that measurement of 18:2(9-*cis*,11-*trans*) by HPLC might become a useful tool in diagnosis of cervical precancer. Subsequently, however, Domandy et al. acknowledged that certain bacteria in the vaginal flora, including *Corynebacterium* sp and *Lactobacillus* sp, could generate 18:2(9-*cis*,11-*trans*) from 18:2(9,12-*cis*), and they withdrew their claim that measurement of the diene might be a possible alternative screening test for cervical intraepithelial neoplasia (Fairbank et al., 1988). Nevertheless, methods for the measurement of the isomer 18:2(9-*cis*,11-*trans*) in serum (Iversen et al., 1985) and tissue samples (Griffin et al., 1987; Tay et al., 1987) are described.

5.1.3.1. Method
REAGENTS
0.1 M Tris (hydroxymethyl)aminomethane buffer (pH 8.9)
Methanol (HPLC grade)
Acctonitrilc,
Propan-2-ol   } S-grade from Rathburns Chemicals
Acetic acid   } Peebleshire, Scotland.
Cholesterol esterase, 10 000 U/l; lipase 20 000 U/l and phospholipase A_2 (from *Naja naja* venom: 10 000 U/l; Sigma Chemicals, U.K.).
β-Eleostearic acid (internal standard [18:3(9,11,13-*cis*)]) from Alltech Associates, U.K.
Octadeca-9,11-*trans*-11-dienoic acid (supplies are obtainable from Dr. D.G. Wickens, Department of Chemical Pathology, Whittington Hospital, London, N19 5NF, U.K.)
'Bond Elut' or 'Sep-Pak' octadecyl (C18) disposable sample preparation cartridges (from Jones Chromatography Llanbradach, Glamorgan, U.K., and Wates Associates).

DETAILS OF PROCEDURE
1. Serum or plasma (0.5 ml) is hydrolysed by mixing with 0.5 ml of

a solution containing 0.1 M Tris buffer (pH 8.9), 1 M methanol and 5000 U/l phospholipase A_2. Alternatively, larger volumes of serum (2 ml) can be hydrolysed by incubation with 2 ml of Tris buffer (0.1 M, pH 7.7) containing phospholipase A_2 (10 000 U/l), cholesterol esterase (10 000 U/l), lipase (20 000 U/l) and methanol (1 M) (Iversen et al., 1984). Tissue samples (4–10 mg) can be homogenized in 0.2 ml of 100 mM Tris buffer (pH 8.9) containing methanol and phospholipase A_2 (Griffen et al., 1987) (These enzyme solutions are stable at 4°C for up to 3 weeks).

2. After incubation at 25°C for 15 min, 2 vol. of methanol are added containing 0.5% (v/v) acetic acid and the internal standard, β-eleostearic acid, (50 mg/l) or octadeca-9,11-trans-dienoic acid (10 μM). The internal standards are stable for 3 months at −20°C when protected from the light.

3. After centrifugation at 1000 × g for 10 min, a known volume of the supernatant is applied to a 'Bond-Glut' or 'Sep-Pak' cartridge. This is pre-conditioned by washing twice with 2.5 ml of a solution of methanol/water/acetic acid (67:33:0.04) immediately before use. Samples are eluted with 1 ml of propan-2-ol/acetonitrile (2:1,v/v).

4. The eluate is analysed by HPLC using a Spherisorb ODS2 reversed-phase packing (250 × 4 mm) containing 5 μm fully capped spherical particles (Hichrom, Reading, U.K.) and a mobile phase of acetonitrile/water/acetic acid (85:15:0.1) delivered at 1.5 ml/min. Samples are injected via a Rheodyne (Cotati, California, U.S.A.) syringe-loading value injector fitted with a 50 μl loop. Conjugated dienes and the internal standards are detected at 234 nm and unconjugated fatty acids at 210 nm by utilizing two variable wavelength UV detectors in series (Fig. 5.2).

5. *Calibration*: Standard curves are prepared for added linoleic acid, linolenic acid, arachidonic acid and internal standards. Results are calculated using the ratio of the peak-height of the 18:2(9-*cis*,11-*trans*) biological isomer to that of the internal standard and multiplying by the concentration of the internal standard added.

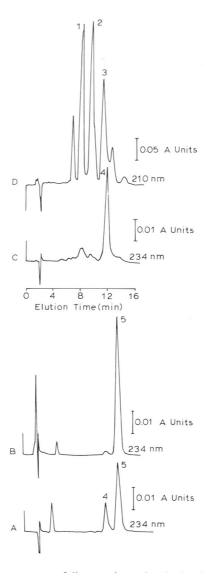

Fig. 5.2. HPLC chromatograms of diene conjugated and related fatty acids. **(A)**: UV-irradiated linoleic acid with albumin; **(B)**: the synthetic isomer of 18:2(9,11)diene; **(C)** and **(D)**: hydrolysed serum at 234 and 210 nm; Peak 1: arachidonic acid; Peak 2: lino-lenic acid; Peak 3: linoleic acid; Peak 4: biological isomer of 18:2(9,11)diene; Peak 5: synthetic isomer of 18:2(9,11)diene. (Iversen et al., 1984.)

5.1.4. Measurement of loss of polyunsaturated fatty acids during lipid peroxidation

Measurement of the selective loss of polyunsaturated fatty acids from membrane fractions can be used to assess the degree of lipid peroxidation in vitro where there is frequently a good correlation with other indices of peroxidation such as the TBA test (Jordan and Schenkman, 1982; Smith and Anderson, 1987) (q.v., Section 3.1.2 in this Chapter). When used to assess peroxidation in vivo, difficulties arise if attempts are made to use this method to assess peroxidation in vivo because the measurement of small differences between very large numbers, and also the mechanisms that operate for fatty acid replacement, can be expected to mask prior peroxidative changes (Buttriss and Diplock, 1988).

Fatty acids are usually measured as their methylesters by gas liquid chromatography on packed, capillary or surface-coated open tubular (SCOT), columns with flame ionization detection (Christie, 1982a). HPLC can also be used to separate fatty acid esters with measurement by UV-absorption or fluorescence detection (Christie, 1987).

5.1.4.1. Measurement of glycerophospholipid fatty acids by gas liquid chromatography (GLC)

This procedure may be divided into four stages:
1. extraction of lipids
2. TLC to separate the glycerophospholipids from other lipids
3. esterification of the glycerophospholipid fatty acids
4. GLC separation of the resultant fatty acid methyl esters

5.1.4.1.1. Extraction Exhaustive tissue extraction procedures are usually employed to isolate glycerolipids from biological matrices. Quantitative results can be achieved using the method of Bligh and Dyer (1959).

REAGENTS
Chloroform/methanol (1:2, v/v)

Potassium chloride (2 M) in 0.5 M phosphate buffer (pH 7.2)
Water/chloroform/methanol (2:5:5, v/v/v)

DETAILS OF PROCEDURE

Extraction. To 1 ml of tissue homogenate is added 3.75 ml of chloro-
form/methanol solution so that the ratio of chloroform/methanol/
water in the mixture is 5:10:4 (v/v/v). After thorough mixing, the mixture is converted to a biphasic system by addition of 1.25 ml of chloroform and a similar volume of potassium chloride reagent so that the
ratio of components is 10:10:9 (v/v/v). After centrifugation at 2000 × g
for 10 min, the upper phase is carefully aspirated so that the precipitate which forms at the interphase is undisturbed and a further 4.58
ml of the water: chloroform/methanol mixture is added. After mixing,
the phases are separated by centrifugation and the lower phase collected quantitatively into a clean stoppered test-tube and evaporated
under N_2 at 20°C.

5.1.4.1.2. Thin-layer chromatography (TLC) of phospholipids   Phospholipids can be purified from the lipid extract by TLC on 0.5 mm
layers of Kieselguhr H (Merck, BDH Chemicals) using a solvent system of hexane/diethylether/glacial acetic acid (60:30:1, v/v/v) (Lynch
et al., 1976). The phospholipids, which remain at the origin, are
scraped from the silica into 2 ml of chloroform/methanol/water
(5:5:1) (Christie, 1982a). Quantitative recoveries can be achieved by
repeating this washing process at least twice and the combined solvents are evaporated under reduced pressure at 37°C. Numerous TLC
methods exist for separation of individual phospholipid classes
(Christie, 1982a).

5.1.4.1.3. Methylation of fatty acids   Glycerophospholipids can be
transesterified directly by basic and acidic procedures without need
for saponification of fatty acids.
(a) Alkaline transesterification (Christie, 1982b)
     O-Acyl lipids are transesterified rapidly in anhydrous methanol in

the presence of a basic catalyst. Free fatty acids are not esterified and all water must be removed prior to esterification.

REAGENTS
Diethyl ether (stored over sodium wire)
Methyl acetate
1 M Sodium methoxide in methanol (stable for several months at room temperature)
Glacial acetic acid
Cyclohexane

PROCEDURE
To 1–10 mg of glycerophospholipid, add 1 ml diethyl ether, 25 μl of 1 M sodium methoxide and 25 μl of methyl acetate. After 5 min, 6 μl of acetic acid is added, the solution centrifuged at 1500 × g for 2 min and the solvent carefully evaporated under N_2. The residue is re-dissolved in cyclohexane before injection.

(b) Use of boron trifluoride (Morrison and Smith, 1964)
Boron trifluoride in methanol is a rapid and simple transesterification catalyst but it has a limited shelf-life unless refrigerated; the use of old solutions that are too concentrated can result in the production of artifacts and the loss of large amounts of polyunsaturated fatty acids.

REAGENTS
Boron trifluoride in methanol (14%, w/v)
Pentane

PROCEDURE
Glycerophospholipids are transesterified by addition of 1 ml reagent which is gassed with N_2 and heated for 10 min in a boiling-water bath. The methylesters are then partitioned into 2 ml of pentane by addition of 1 ml of water, and the upper phase collected and evaporated under N_2.

Internal standardization: Quantitative measurements may be made by an internal standard method using heptadecanoic acid which can be added to the samples before transesterification. The amount added is dependent on the concentration of free fatty acids in the sample.

5.1.4.1.4. Gas liquid chromatography   Fatty acids esters can be separated on the basis of molecular weight on packed columns with silicone liquid phases such as SE-30, OV-1, JXR or QF-1. Saturated and unsaturated fatty acids of the same chain length can be separated at low levels of packing ($\sim 1.3\%$), but polar polyester liquid phases allow complete separation of esters of the same chain-length with zero to six double bonds. Components are resolved in order of the degree of saturation, unsaturated components being eluted after the relatively saturated ones. High polarity phases such as polymeric ethyleneglycol succinate (EGS), di(ethylene glycol) succinate (DEGS) and medium polarity phases such as poly(ethylene glycol) adipate (PEGA) and EGSSY (a copolymer of EGS) are the most common phases in packed columns.

PROCEDURE (using 10% diethyleneglycol succinate (DEGS))

Column	:	10% DEGS, 100/120 mesh, Diatomite C
Size	:	1.5 m × 4 mm
Oven temperature	:	190°C
Detection	:	Flame ionization at 250°C
Injection temperature	:	200°C
Carrier gas	:	N_2 at 1.3 kg/cm^2
Analysis time	:	40 min

More recently, capillary or open tubular columns have in part replaced packed columns in GLC analysis because of their high efficiency (100 000 theoretical plates overall) and their high resolving power. A number of methods for the separation of fatty acid methylesters have been reviewed (Christie, 1982a).

5.2. Propagation processes and lipid hydroperoxide formation

5.2.1. Introduction

Radicals (R·) formed from conjugated lipid dienes react spontaneously with oxygen to form peroxy radicals:

$$R· + O_2 \rightarrow ROO·$$

Peroxy radicals are either quenched by antioxidants or react with adjacent polyunsaturated lipids by abstraction of hydrogen, thus forming lipid hydroperoxides (ROOH):

$$ROO· + RH \rightarrow R· + ROOH$$

The measurement of lipid hydroperoxides in biological samples is frequently used as an index of lipid peroxidation. In practice such measurements are complicated by the lability of these compounds in the presence of metal ions, and thiols and by the metabolism of hydroperoxides by peroxidases such as glutathione peroxidase and the glutathione S-transferases.

5.2.2. Measurement of lipid hydroperoxides by iodometric determination

5.2.2.1. Spectrophotometric measurement
In the absence of oxygen, peroxides react with iodide in a 1:1 stoichiometric relationship and liberate iodine (Lea,1931;1952):

$$ROOH + 2H^+ + 2I^- \rightarrow ROH + H_2O + I_2$$

The iodine liberated can be estimated by thiosulphate titration, or spectrophotometry can be used to measure the formation of the tri-iodine anion (I_3^-) at 353 nm, which is produced in the presence of excess iodide (Pryor and Castle, 1984; Chan and Coxon, 1987).

$$I_2 + I^- \rightleftharpoons I_3$$

Two procedures have been adopted to prevent oxidation of excess unreacted iodide by molecular oxygen. Either unreacted iodide is complexed with cadmium (Buege and Aust, 1978), or spectrophotometric measurements are made in air-free stoppered cuvettes and development of I_3^- is followed continuously. The second procedure allows reliable detection of the reaction end point and identifies any oxygen contamination (Hicks and Gebicki, 1979), but it is not applicable to routine measurements of many samples as the time taken for complete reduction of the hydroperoxide is prohibitive. Methods are therefore described which are based only on the procedure utilizing cadmium.

REAGENTS

Chloroform/methanol 2:1 (v/v)

Chloroform/acetic acid 2:3 (v/v).

These reagents are deoxygenated by bubbling with nitrogen at 4°C, then sealed and allowed to come to room temperature.

Cadmium acetate, 0.5% (w/v)

Potassium iodide, 120% (w/v); prepared freshly within 15 min of use by dissolving 6 g of potassium iodide in 5.0 ml of water at 4°C. (The water is previously bubbled with nitrogen for 15 min.)

Cumene hydroperoxide

PROCEDURE

Lipids are extracted from 1 vol. of membrane suspension in a stoppered test-tube by shaking with 5 vol. chloroform/methanol reagent. The sample is centrifuged at $7000 \times g$ for 5–10 min, and the upper phase removed by aspiration. A known volume of the lower chloroform phase (3 ml) is recovered and taken to dryness under nitrogen in a water bath at 45°C. While the lipid residue is still under the nitrogen, 1.0 ml of chloroform/acetic acid reagent and 0.05 ml potassium iodide (120%, w/v) are added, the test-tube stoppered, and the contents thoroughly mixed. The sample is placed in the dark for exactly 5 min after which 3.0 ml of cadmium acetate solution is added. The solution is again mixed and centrifuged at $1000 \times g$ for 10 min. The absorbance of the upper phase is measured at 353 nm against a blank containing all reagents except the sample.

CALIBRATION
The assay can be standardized by using a solution of the stable compound in chloroform/methanol, and relating the amount of I_3^- liberated to the known concentration of cumene hydroperoxide.

The molar absorption coefficient of I_3^- in chloroform/acetic acid reagent is

$$1.73 \times 10^4 \text{ M}^{-1} \cdot \text{cm}^{-1}$$

The sensitivity of the assay is in the range 0.05–0.5 μmol of hydroperoxide.

5.2.2.2. *Assay of hydroperoxides by the iodometric technique, with HPLC measurement of I_3^-*
Gcbicki and Guille (1989) improved the sensitivity of the tri-iodide spectrophotometric assay by an HPLC method for measuring I_3^-. The detection limit is 100 pmol of peroxide, and the assay is linear over a range from 100 pmol to 25 nmol.

REAGENTS AND PROCEDURE
HPLC grade reagents (Rathburns Chemicals) are used throughout. The equipment used to prepare samples for assay consists of a line supplying helium passing via a presaturator filled with methanol/acetic acid (2:1, v/v). The gas stream is then split so that one part passes through a regulating valve into a distributor with six outlets made of stainless steel tubing (Pierce Chemical Co.), while the remainder enters an Oxford pipettor vessel where it emerges through a stainless

steel sinter. Liquid in the pipettor is mixed with a magnetic stirrer (Fig. 5.3). A mixture of methanol/acetic acid (2:1, v/v) is added to the pipettor and the solvent degassed with helium for 20 min. Solid potassium iodide is then added to a concentration of 1% (w/v) and dissolved by stirring. Up to six glass vials, each stoppered with Mininert valves (Pierce Chemical Co.) are then attached to the stainless steel part of the distributor and gassed for 3 min with helium, using 21-gauge syringe needles fitted loosely through the valve capillary opening. After the gas delivery tubes have been withdrawn, the aqueous or organic hydroperoxide extracts obtained as detailed in Section 5.2.2.1 are added to the vials using a microsyringe, and 2 vol. of methanolic potassium iodide solution is added from the pipettor. The vials are closed off and incubated at 50°C for 15–20 min; finally, 0.1 vol. of cadmium acetate (8%, w/v) in methanol/water (1:1, v/v) for every volume of iodide used, is then added to each vial. After mixing, the samples are analysed by HPLC.

Measurement of I_3^- absorbance is carried out at 358 nm. The

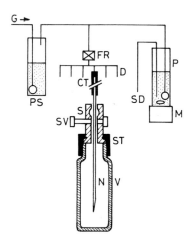

Fig. 5.3. Apparatus used in the iodometric assay. G: inert gas supply; PS: presaturator; P: pipettor; M: magnetic stirrer; SD: solvent delivery nozzle; FR: flow regulator; D: gas distributor; CT: connecting tube; V: vial; S: stopper with capillary opening; SV: side valve; ST: screw-top stopper holder; N: syringe needle. (Gebicki and Guille, 1989.)

anion is resolved on a C-18 reversed-phase HPLC column (Waters Associates) using a mobile phase of methanol/water/acetic acid (2:2:1, v/v/v) at a flow rate of 2.0 ml/min. Between 5–100 μl of sample can be injected, and the method standardized by comparison of sample peak areas with known amounts of standard prepared by reaction of known amounts of H_2O_2 under similar conditions.

5.2.2.3. Enzymatic assay of lipid hydroperoxides
A selective assay has been described for lipid hydroperoxides, based on the cyclooxygenase reaction with a sensitivity in the range of 20–200 pmol (Marshall et al., 1985; Pendleton and Lands, 1987; Lands, 1988). The substrate (15-HPETE) is not, however, commercially available and must therefore be synthesized from purified arachidonic acid by the method of Graft (1982) in which soya bean lipoxygenase is used to catalyse the stereospecific oxygenation of arachidonic acid.

5.2.2.3.1. Synthesis of 15-hydroperoxy-5,8,4,13-eicosatetraenoic acid
        (15-HPETE)
REAGENTS
5,8,11,14-[1–14C]eicosatetraenoic acid (Amersham)
5,8,11,14-eicosatetraenoic acid
12-Hydroxyoctadecanoic
12-Hydroxy-9-octadecanoic acids
Soybean lipoxygenase, type 1, 170,000 units/mg protein (Sigma Chemicals)
100 mM Tris-HCl buffer (pH 9.0)
0.23 N hydrochloric acid
Ethylacetate (Spectral grade)
Petroleum ether (b.p. 35–60°C) shaken gently with 1/10 vol of concentrated sulphuric acid, washed with distilled water and distilled over lithium aluminium hydride.
Silicic acid, 100 mesh (Slurry 500 g five-times each with 5 litres of water. Discard the supernatant after 5 min and activate for 12 h at 150°C).
Tris(trimethylsilyl)trifluoroacetamide.

Purification of arachidonic acid (5,8,11,14-eicosatetraenoic acid)
Peroxides are removed from commercially available arachidonic acid
by addition of 15 mg of sodium borohydride to 100 mg fatty acid dissolved in 3 ml toluene. The mixture is incubated for 30 min with occasional stirring before addition of 3 ml water and 0.7 ml 1 M citric acid.
After vortex-mixing, the phases are separated and the organic phase
removed and washed with 1 ml of water. Peroxide free arachidonic
acid is dried over anhydrous sodium sulphate and stored at 4°C after
addition of 2 ml of 10 mM BHT.

PROCEDURE
The radiolabelled 5,8,11,14-[1-^{14}C]eicosatetraenoic acid (5 μCi) is diluted with 40 μmol unlabelled fatty acid to a final specific activity of
250 cpm/nmol. The solvent is evaporated under N_2 and the eicosatetraenoic acid dispersed in 2 ml 100 nM Tris-HCl buffer by sonication
for 2 min at 20°C. The fatty acid suspension is subsequently transferred into a 250 ml flask containing 47 ml 100 nM Tris-HCl preequilibrated at 30°C. Hydroperoxides are then prepared from the fatty acid
by addition of 20 mg of soybean lipoxygenase dissolved in 1.0 ml 100
mM Tris-HCl buffer. The solution is incubated for 5 min at 30°C with
constant shaking in an orbital shaker at 200 rpm and gassed with oxygen at a rate of 30 ml/min. The reaction is terminated, and the lipoxygenase product extracted by transferring the mixture to a separating
funnel containing 100 ml of petroleum ether/ethylacetate (1:1, v/v)
precooled to −25°C and 20 ml of 0.23 N HCl. The organic phase is
separated, and the aqueous phase re-extracted twice with 100 ml of
solvent. The combined solvent is stored in a desiccator at −7°C over
anhydrous sodium sulphate. After 20 min, the solids are removed by
vacuum filtration and the filtrate concentrated to near dryness
($\approx$ 100–200 μl) before addition of 2 ml 3% ethyl acetate. This procedure is repeated three times before the sample is finally dissolved in
1.0 ml 3% ethyl acetate. The 15-L-hydroperoxy-5,8,11,13-eicosate-
traenoic acid (15-HPETE) is then purified on silicic acid. A slurry of
3 g of silicic acid in 3% ethyl acetate in petroleum ether is poured into
a 1 × 10 cm glass column and the sample applied, and eluted at 0.8

ml/min with 80 ml 3% ethyl acctate followed by 160 ml 8% ethyl acetate in petroleum ether and finally 50 ml of ethyl acetate. 15-HPETE is eluted from the column with 8% ethyl acetate/petroleum ether in a column volume of 100–130 ml. The solvent is evaporated to a volume of 100–200 μl and 15-HPETE further purified by TLC using 0.25 mm $\times$ 10 $\times$ 20 cm Silicagel 60H plates and a mobile phase of diethyl ether petroleum ether acetic acid 50:50:0.5 (v/v/v) preequilibrated at room temperature for 20 min. The TLC plates are developed to a distance of 13.5 cm above the origin, a portion of the plate sprayed with a 0.1% ethanolic solution of 2,7-dichlorofluorescein and visualized under UV light; 15-HPETE (R_F value 0.36) is then recovered from the plate. It is important to minimalize decomposition of extracted 15-HPETE by reducing the temperature of the extract immediately to $-20°$C. Procedures should be carried out in a cold room wherever possible.

5.2.2.3.2. Enzymatic assay of hydroperoxides in plasma

REAGENTS

15-HPETE

Peroxide-free arachidonic acid

Haemin solution

Butylated hydroxytoluene (10 mM)

Cyclooxygenase (stored at $-70°$C)

Potassium phosphate buffer (pH 7.2, 0.1 M)

Phenol

Sodium cyanide

PROCEDURE

Platelet-poor plasma is prepared from venous blood by centrifugation at $1000 \times g$ for 15 min, and an equal volume of ethanol is added. The mixture is left on ice for 10 min and then centrifuged at $6500 \times g$ for 15 min. Cyclooxygenase activity is measured polarographically, using an oxygen electrode and an incubation buffer (total volume 3 ml) containing potassium phosphate (pH 7.2, 0.1 M), 100 μM arachidonic acid, 1 mM phenol and 2.5 mM sodium cyanide. The reaction is started by addition of 50–125 μl of sample and 5 μg of cyclooxygenase holoenzyme (reconstituted in 5 μl of haemin solution (1 μg/μl)). The system is gassed with 70 μmol of oxygen/min per μg protein, and the time taken to reach optimal velocity (the 'lag' time) is recorded.

Calculation of the hydroperoxide content

Measurements of the 'lag' time are made in the absence of exogenous hydroperoxide in a cyanide-treated system (Lag_{CN}), a control system not including cyanide ($\text{Lag}_{\text{Control}}$), and a cyanide-inhibited system to which hydroperoxide had been added ($\text{Lag}_{(\text{CN}+\text{ROOH})}$). A 'fractional activation' of the enzyme response to the added hydroperoxide is calculated as:

$$\frac{\text{Lag}_{\text{CN}} - \text{Lag}_{(\text{CN}+\text{ROOH})}}{\text{Lag}_{\text{CN}} - \text{Lag}_{\text{control}}}$$

This allows normalization of responses so that results from different experiments can be compared.

A standard curve of 'fractional activation' against the concentration of pure 15-HPETE is also prepared, and the concentration in plasma calculated by reference to this calibration curve.

The advantage of this method lies in its sensitivity (the limit of detection is 6 pmol/assay) but the disadvantage is the need for a purified lipid hydroperoxide standard which is a time consuming and difficult preparation.

5.2.2.4. *Analysis of specific lipid hydroperoxides by HPLC*

Fatty acids hydroperoxides can be measured by HPLC. This not only allows quantitation of specific products of lipid peroxidation but allows recognition of product distributions characteristic of radical-initiated oxidation, singlet oxygen reactions and the product of specific cyclooxygenase, lipooxygenase and *P*-450 enzyme systems (Smith and Anderson, 1987). A schematic diagram (Scheme 5.1) for the analysis of blood plasma is given below, derived from Frei et al. (1988).

Free fatty acid hydroperoxides and phospholipid hydroperoxides partition into the aqueous methanol layers, whereas triglyceride hydroperoxides and cholesterol ester hydroperoxides are recovered in

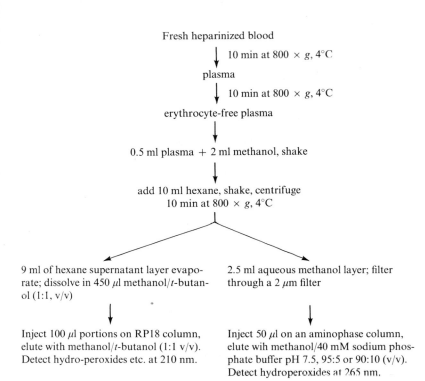

Scheme 5.1. Analysis of blood plasma.

the non-polar solvent (Yamamoto and Ames, 1987; Frei et al., 1988). The method has been applied to human and rat tissues, and a good correlation was found between total lipid hydroperoxides and conjugated diene levels in rat liver with or without dietary modifications of vitamin E content (Slater, 1988).

5.3. Aldehyde breakdown products of lipid peroxidation

5.3.1. Malondialdehyde

5.3.1.1. Introduction
Malondialdehyde (MDA) is a highly reactive three carbon/dialdehyde

produced from lipid hydroperoxides. It can, however, also be derived by the hydrolysis of pentoses, deoxyribose, hexoses, from some amino acids and from DNA (Pryor et al., 1976; Frankel and Neff, 1983). MDA has most frequently been measured by the thiobarbituric acid reaction (TBA test) (Kohn and Liversedge, 1944), but there are now highly specific and accurate HPLC methods which distinguish between true MDA and other aldehydes that may react with TBA.

5.3.1.2. The thiobarbituric acid test

One molecule of MDA reacts stoichiometrically with two molecules of 2-thiobarbituric acid (Sinnhurber et al., 1958; Yu et al., 1986); the reaction occurs at a pH of 2–3 but excess acid (pH < 2) inhibits the colour development.

The adduct (Fig. 5.4) formed exhibits a strong primary maximum at 532–535 nm and weaker secondary maxima at 245 and 305 nm. The pink chromogen can be detected spectrophotometrically and has a molar absorption co-efficient between 149 000 litre $\cdot$ mol^{-1} $\cdot$ cm^{-1} and 156 000 litre $\cdot$ mol^{-1} $\cdot$ cm^{-1} (Kohn and Liversedge, 1944; Slater and Sawyer, 1971; Nair and Turner, 1984). The adduct also exhibits fluorescence at an excitation wavelength of 515–532 nm and emission wavelength of 553 nm (Sinnhurber and Yu, 1958; Jordan and Schenkman, 1982). The measurement of TBA reactivity in complex biological material is due to MDA (and other aldehydic products) derived from non-volatile primary peroxidation products, which decompose to form MDA during the acid-heating stage of the assay (Jordan and Schenkman, 1982; Fraga et al., 1988). A wide variety of chemical species such as, saturated and unsaturated aldehydes, substituted pyrimidines, sulphadiazine, biliverdin, sucrose, fructose, glucose, 2-deoxyribose and N-acetylneuraminic acid have also been shown to react with

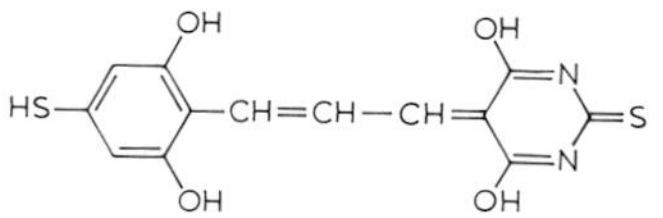

Fig. 5.4. Thiobarbituric acid adduct with malondialdehyde.

TBA, thus interfering with the assay. Haemolysed red cells also cause an apparent increase in serum MDA levels (Gutteridge and Tickner, 1978). Saturated alkanals react with TBA under normal assay conditions to produce yellow and red pigments with very weak absorptions at 455 and 532 nm so their contribution to the total is small. Unsaturated aldehydes produce adducts with prominent orange chromogens which absorb strongly at 495 nm and to a lesser degree at 525–535 nm. In the presence of sugars, however, even unsaturated aldehydes react with TBA to produce a chromogen with an absorption maxima sequence of 532 > 500 > 445 nm. Direct measurement by HPLC has confirmed that the 532 nm absorbance is due to MDA. Thus sugars and haemoproteins in the presence of transition metals strongly affect the TBA reaction and can lead to misleading values that are attributed to increased lipid peroxidation (Gutteridge and Tickner, 1978). The addition of anti-oxidants such as 0.01% BHT to the TBA reagent lowers the metal-catalysed auto-oxidation of lipids. Lowering the reaction temperature in the heating-step to 80°C has been shown to reduce interference from sucrose present in buffers (Yagi, 1984). Good correlations between the TBA reaction and MDA levels measured by HPLC can be achieved when protein-free supernatant fractions are assayed following acid precipitation (Cheeseman et al., 1988).

TBA reactivity is thus a reflection of the total amount of MDA, β-unsaturated aldehydes, cyclic peroxides and 'contaminants' present in the sample and in the basic form, the reaction is clearly not specific for any one class of peroxidation product (Kosugi and Kugawa, 1986). However, as a first approximation, measurement of tissue fraction oxidation it can be a useful method because it is quick and easy. Great care must be exercised in the interpretation of the results.

5.3.1.3. Methods for the TBA-reaction

5.3.1.3.1. (a) Using colorimetry at 535 nm: preparation of MDA MDA is prepared from the acetal by acid hydrolysis. A solution containing 0.1 mmol 1,1,3,3-tetraethoxy- or methoxypropane in 50 ml water and 0.1 ml, 0.1 M HCl is warmed at 50°C for 1 hour and the volume adjusted to 100 ml with water. The concentration of free MDA is determined spectrophotometrically at 267 nm, using a molar absorption co-efficient of 31 800 (Kwon and Watts, 1963) and a calibration line of MDA concentration against TBA-reactivity is prepared.

METHOD (Buege and Aust, 1978)

REAGENTS
Trichloracetic acid (15% (w/v) in 0.25 N HCl) ⎱
Thiobarbituric acid (0.37% (w/v) in 0.25 N HCl) ⎰ stock reagent

PROCEDURE
One volume of sample containing 0.1–2.0 mg of protein or 0.1–0.2 μmol of membrane lipid phosphate and 2.0 vol of stock reagent are combined in a screw-capped centrifuge tube, mixed and heated for 15 min in boiling water. After cooling, the precipitate is removed by centrifugation at $1000 \times g$ for 10 min and the absorbance of the supernatant measured at 535 nm against a sample-blank containing reagents but no sample. Quantitation can be obtained using a molar absorption co-efficient of $156\,000$ M^{-1} · cm^{-1} or by using a calibration curve prepared with MDA and the TBA reagent.

This method has been improved by:
1. extracting the chromogen in n-butanol;
2. by the use of double wavelength measurements to increase the assay specificity; and
3. by preparation of a standard curve using free MDA (Uchiyama and Mihara, 1978).

REAGENTS
Phosphoric acid (1% (w/v) in 0.1 N HCl)
Thiobarbituric acid (0.6% (w/v) in 0.1 N HCl)
n-Butanol (Analar grade)

PROCEDURE
One volume of sample or standard, 6 vol of phosphoric acid and 2

vol of TBA are combined in a stoppered test-tube, mixed and heated for 45 min in boiling water. After cooling, the chromogen is extracted in 8 vol of n-butanol. (Dilute preparations can be concentrated by evaporation of the solvent at 40°C under partial vacuum.) The absorbance of the organic phase is determined at 535 and 520 nm against a sample blank, and the difference in absorbance used as the 'TBA value' by relating it to an appropriate MDA standard.

5.3.1.3.2. (b) Using fluorescence measurements   A method for measurement of lipid peroxides in plasma and serum (Yagi, 1984), that overcomes interference from water-soluble substances that may form adducts with TBA involves the use of a spectrofluorimetric technique.

REAGENTS
Saline (0.9% w/v)
N/12 sulphuric acid
Phosphotungstic acid (10%, w/v)
Thiobarbituric acid (0.67%, w/v) dissolved in 50% (w/v) glacial acetic
   acid
n-Butanol

PROCEDURE
One volume of whole blood ($\sim$ 50 μl) and 20 vol of saline are mixed and centrifuged at 3000 $\times$ g for 10 min. Either 0.5 ml of supernatant fluid or 20 μl fresh serum are combined with 4.0 ml of sulphuric acid and 0.5 ml of phosphotungstic acid. After 5 min, the mixture is centrifuged and the supernatant discarded. Cell lipids and proteins are resuspended in 2.0 ml of sulphuric acid and 0.3 ml of phosphotungstic acid and re-centrifuged. The sediment is resuspended in 4.0 ml of water, and 1.0 ml of TBA reagent added. The samples are heated at 95°C for 60 min and thiobarbituric acid reactive substances (TBARs) extracted with 5.0 ml of n-butanol. Fluorescence measurements are made at excitation and emission wavelengths of 515 and 555 nm, respectively, and the method standardized against MDA.

5.3.1.4. Determination of MDA-TBA adduct by HPLC

To avoid spurious absorbance at 532 nm arising from adduct formation of TBA with complexes other than MDA, several methods for the separation of the TBA-MDA complex by reversed-phase HPLC have been developed (Bird et al., 1983; Bird and Draper, 1984; Gilbert et al., 1984; Yu et al., 1986). The methods are very sensitive and specific but because hot acidic conditions are necessary to obtain the adduct there is the possibility of generating MDA artificially.

PROCEDURE (Bird et al., 1983; Bird and Draper, 1984)

Samples are homogenized in 5% TCA, and centrifuged at $1000 \times g$ for 10 min. Adduct is formed with 1% aqueous TBA at 100°C for 30 min, and the pH of the sample is adjusted to pH 1.5 with 4 M HCl and the sample re-centrifuged. An aliquot of the supernatant fluid is neutralized with 0.15 M NaOH in distilled water/methanol (1:1, v/v) and 10–20 μl then injected onto the column (see below). Alternatively, the adduct may be extracted in n-butanol, the solvent evaporated, and the residue redissolved in mobile phase for injection. The detection limit is about 250 pg of TBA-MDA adduct.

HPLC CONDITIONS

Column	:	μBondapak C_{18} with C_{18} Corasil guard column
Column dimensions	:	0.39×30 cm with 0.3×2.2 cm guard
Mobile phase	:	Methanol/water (11:89, v/v)
Flow rate	:	2.0 ml/min
Temperature	:	ambient
Detection	:	535 nm
Sample injection	:	10–20 μl
Analysis times	:	6–8 min
Sensitivity	:	1 ng

Note: The column should be washed with about 100 ml of methanol after 15–20 determinations.

Yu et al. (1986) achieved a detection limit of 3.8 pg for the TBA-MDA adduct using fluorometric detection (532 nm excitation; 550 nm emission). The adduct was separated on a Zorbax C18, 3 μm column (0.63 × 15 cm) or an Altex ODS, 5 μm analytical column (0.46 × 25 cm); the columns were eluted with 0.01–0.025 M phosphate buffer (pH 6.5)/methanol (60:40, v/v) at a flow rate of 1.0 ml/min. A presaturation column packed with Corasil silica (35–50 μm) was placed between the pump and the column. Reducing the pH of the mobile phase had little effect on retention time but symmetry and band broadening were worse at more acidic pH probably because the thionyl group of the TBA adduct is increasingly ionized at pH 6.5 and is less attracted to similarly charged residual silanols on the ODS columns.

5.3.1.5. Determination of free MDA by HPLC

Below a pH of 4.65, free MDA exists in an undissociated cyclic form ($\varepsilon_{245} = 30\,000$). Although it can be determined by UV-absorptiometry (Kwon and Watts, 1963; Kukuda et al., 1981) and differential pulse polarography, HPLC techniques are far superior, because they are sensitive and specific.

A number of HPLC methods for the detection of free MDA have been reported which include a size-exclusion method, a method that determined MDA-protein adducts, and a method that involves the determination of the MDA anion.

5.3.1.5.1. Size-exclusion HPLC

Csallany and co-workers separated free MDA in biological samples on a TSK G1000 PW column (Csallany et al., 1984; Lee and Csallany, 1987) and MDA containing proteins on a TSK G2000 SW column (Mainwaring and Csallany, 1988). We have used these methods for the determination of free MDA in tissues of vitamin E-deficient rats at 'endogenous levels' (Csallany, 1984; Tokarz et al., 1988). The disadvantage of the method is the long run-time needed for individual samples.

PREPARATION OF SAMPLES FOR HPLC

Tissue samples (0.2–1 g) are weighed and homogenized for 15 s in

10–15 vol of 0.01 M Na_3PO_4 buffer (pH 8.0) using a Polytron homogenizer. It is necessary to include an antioxidant in the homogenate (100 μg α-tocopherol) to avoid generation of MDA during the homogenization (Tokarz et al., 1988). The homogenate is then centrifuged at $2000 \times g$ for 15 min, and the supernatant fluid filtered in a 50 ml Amicon cell equipped with a UM05 membrane ultrafilter (Amicon Corp, MA, U.S.A.) pressured at 35 psi with nitrogen. This removes compounds larger than 500 kDa and at least 2 ml of supernatant fluid is required for reproducible filtration.

HPLC CONDITIONS

Column	:	Spherogel-TSK G1000 PW (Anachem, Luton, U.K.)
Column dimensions	:	0.75×30 cm
Mobile phase	:	0.1 M Sodium phosphate buffer (pH 8.0)
Flow rate	:	0.6 ml/min
Temperature	:	ambient
Detection	:	UV 267 nm
Sample injection	:	20–2000 μl
Analysis times	:	$\sim$ 50 min
Sensitivity	:	1 ng

CALIBRATION

Standard MDA, produced from acid hydrolysis of 1 nmol tetraethoxypropane is dissolved in 90 ml water and 1 ml 1 N HCl added. The volume is brought to 100 ml and the solution warmed in a water bath at 50°C for 60 min. A 1×10^{-4} solution is prepared by dilution of 1 ml stock MDA solution to 100 ml with 0.01 M Na_3PO_4 buffer (pH 7.0). The absorbance of the solution is determined at 267 nm and the concentration calculated assuming a molar absorption co-efficient of 31 506 $M^{-1} \cdot cm^{-1}$. A standard curve in the range 1.5×10^{-7} to 1.2×10^{-6} M can then be derived from peak height measurements. Bound MDA (Lee and Csallany, 1987) can be determined after alkaline saponification of tissue homogenates in sucrose-free buffered solution. An homogenate derived from about 6 g tissue is transferred to a stoppered centrifuge tube, and the pH of the homogenate adjusted to 13, using a saturated solution of NaOH. The sample is incubated at 60°C in a water bath for 30 min, neutralized to pH 8.0 with concentrated HCl and then filtered using the Amicon cell as described above. The filtrate is then analysed by HPLC and the MDA value referred to as the total MDA content of the sample. Bound MDA can be estimated by the difference between levels before and after hydrolysis.

5.3.1.5.2. Analysis of MDA: protein adducts by HPLC   Malonaldehyde-containing proteins can be determined by a combination of open column chromatography using Sephadex G-15 or G-25 (Pharmacia Fine Chemicals) and HPLC (Manwaring and Csallany, 1988). Portions of tissue (0.05–1.5 g) are homogenized in 20 ml chloroform/methanol 2:1 (v/v) using a Polytron stainless steel homogenizer. The homogenate is washed twice with 50 ml water, the water layers are pooled and washed four times with 25 ml 2:1 chloroform/methanol (Manwaring and Csallany, 1981). Water-soluble compounds are then separated on a Sephadex G-15 or G-25 column and the void volume (∼6 ml) collected. Aliquots (100 μl) are injected onto a TSK G2000 SW column and proteins eluted at 1 ml/min in 0.066 M phosphate, 0.3 M sodium chloride, 0.02% sodium azide buffer (pH 7.0) and detected fluorimetrically at 275 nm excitation 350 nm detection (Yost et al., 1977). The approximate molecular mass of the compounds identified are obtained by calibrating the column with protein standards (catalase, 250 kDa; bovine serum albumin, 65 kDa; β-lactoglobulin, 35 kDa; α-chymotoypsinogen A, 23.2 kDa and lysozyme 14.1 kDa). The amount of MDA in each protein peak is determined as previously described.

5.3.1.5.3. Separation of MDA on an amino-bonded silica phase   Separation and measurement of MDA in the enolate anionic form can be achieved using a Waters carbohydrate analysis column (3.5 mm × 30 cm) eluted with acetonitrile/water (Esterbauer and Slater, 1981; Esterbauer et al., 1984). The method is particularly suited to the analysis of MDA generated from peroxidizing tissues in vitro because the analysis time is so short (∼8 min).

PREPARATION OF TISSUES

Aqueous solutions or suspensions of homogenates and subcellular organelles are mixed with an equal volume of acetonitrile and centrifuged at $2000 \times g$ for 10 min. Alternatively, the peroxidation of tissues stressed in vitro during kinetic experiments can be stopped by addition of 0.1 vol desferrioxamine (0.5 mg/ml) and centrifugation of the suspension at $100\,000 \times g$ (Esterbauer et al., 1984). Very lipophilic materials can be removed before HPLC by solid-phase extraction on Waters RP18 Sep-Pak cartridges (Water Associates, Harrow).

HPLC CONDITIONS

Column	:	Spherisorb 55NN amino phase (Hichrom Ltd, U.K.)
Column dimensions	:	0.45 × 20 cm.
Mobile phase	:	acetonitrile: Tris buffer 0.03 M, pH 7.0*
Flow rate	:	1.0 ml/min
Temperature	:	ambient
Detection	:	UV 270 nm
Sample injection	:	20–1000 μl (in mobile phase)
Analysis times	:	4–8 min
Sensitivity	:	1 ng

CALIBRATION

A stock solution of MDA is prepared as described previously, by dilution of 1 mmol to 100 ml with sulphuric acid (1% v/v). The concentration of MDA is determined by UV-spectrophotometry assuming a molar absorption coefficient of 13 700 $M^{-1} \cdot cm^{-1}$ at 245 nm. This solution is then diluted in 0.1 M Tris buffer (pH 7.0) and brought to

---

* The amount of acetonitrile in the buffer may be varied between 30–70%. Optimal conditions should be determined experimentally.

volume with acetonitrile so that the proportions are identical to the mobile phase and the concentration in the range 10–20 μM. Aminophase columns show decreasing retention times with continuous use over a period of weeks. Eventually the retention times stabilize but it is necessary to manipulate the mobile phase by addition of excess acetonitrile to re-optimize the resolution of free MDA. This phenomenon may be due to hydrolysis of the active sites of the column.

5.3.2. Aldehydic breakdown products of lipid peroxidation

5.3.2.1. Introduction
Following peroxidation of n–6 (linoleic, γ-linolenic and arachidonic acids) and n–3 (α-linolenic and docosahexaenoic acids) fatty acids, relatively unstable fatty acid hydroperoxides may be converted by consecutive scission, fission, rearrangement and oxidation reactions into more stable carbonyls. These include: n-alkanals, 2-alkenals, 2,4-alkadienals, alkatrienals, α-hydroxyaldehydes, hydroperoxialdehydes, 4-hydroxyalkenals, 4-hydroperoxyalkenals, MDA, α-dicarbonyls, saturated and unsaturated ketones, alkanes and alkenes (Esterbauer, 1982).

Many studies have characterized the products of lipid peroxidation in hepatic microsomes and other cellular membranes (Benedetti et al., 1980; Esterbauer et al., 1982; Benedetti et al., 1984a,b; Poli et al., 1985; Benedetti et al., 1986; Pompella et al., 1988; Benedetti et al., 1989), blood (Lynch et al., 1983; Ramenghi et al., 1985), tumour cells (Winkler et al., 1984) and lipoproteins (Esterbauer et al., 1987). Most of these studies have followed the development of these products after induction by free radical generating systems in vitro or administration of toxic xenobiotics. Peroxidized biological samples always contain alkanals, 2-alkenals, 2,4-alkadienals and 4-hydroxyalkenals (Esterbauer et al., 1988). The main carbonyls produced during the peroxidation of n–6 fatty acids are hexanal and 4-hydroxy-2,3-*trans*-nonenal, n–3 fatty acids produce propanal and 4-hydroxy-2,3-*trans*-hexenal (Esterbauer, 1982; Esterbauer et al., 1988); all these products are also the most prevalent carbonyls detected in biological tissues. Other aldehydes also detected, but in smaller quantities, are: 4-hydroxy-2,3-octenal, 4-hydroxydecenal, 4-hydroxyundecenal, 4,5-dihydroxydecenal, 4,5-dihydroxyheptenal, 4-hydroperoxynonenal, 4-hydroperoxy-hexenal, 4-hydroxy-2,5-nonenal, butanal, pentanal, hexenal, octenal and nonenal.

It has been suggested that measurement of the aldehydic break-down products of lipid peroxidation is the most reliable indicator that the process has occurred in vivo in normal or pathological conditions (Esterbauer et al., 1988), but there is very little information as to the rate of production of these compounds in vivo (Yoshino et al., 1986; Tomita et al., 1987), or of their relative proportions.

This section is concerned with thin-layer and high-performance liquid chromatographic/mass spectrophotometric techniques used in the analyses of these compounds, which are often found to be present

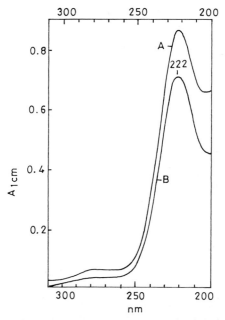

Fig. 5.5. Ultraviolet absorption spectrum of **(A)** biogenic 4-hydroxynonenal and **(B)** synthetic 4-hydroxynonenal; Solvent: acetonitrile/water (1:1,v/v).

TABLE 5.1.

Molar absoption coefficient for free hydroxyalkenals

ε (M$^{-1}\cdot$ cm^{-1})	1 $C_4H_6O_2$	2 $C_5H_8O_2$	3 $C_6H_{10}O_2$	4 $C_7H_{12}O_2$	5 $C_6H_{14}O_2$	6 $C_9H_{16}O_2$	7 $C_{10}H_{18}O_2$
H_2O	15250	13800	13800	13750	13780	13750	13800
Ethanol	13800	13000	13100	13050	13100	13100	13100
Hexane	9750	12200	13000	13850	14750	14400	14200

1. Hydroxybutanal.   2. hydroxypentanal   3. hydroxyhexenal.
4. hydroxyheptenal.   5. hydroxyoctenal.   6. hydroxynonenal.
7. hydroxydecenal.

After Esterbauer and Weger, 1967.

only in the order of a few nanomoles/gram of tissue (Ramenghi et al., 1985; Buffinton et al., 1986; Curzio et al., 1986; Yoshino et al., 1986; Tomita et al., 1987; Esterbauer et al., 1988). As with measurement of conjugated dienes and lipid hydroperoxides, the measurement of aldehydic breakdown products formed as a consequence of peroxidative stress in vivo, represents the balance between the rate of lipid peroxidation and the metabolism of the peroxidized products. Thus failure to detect significant increases in these aldehydes following peroxidative stress can sometimes mask a stimulation of the rate of peroxidation, because the rise in the baseline levels is so small that the capacity of metabolic defences is not exceeded.

5.3.2.2. Detection and characterization of free α,β unsaturated carbonyls

The α,β unsaturated carbonyls exhibit an ultraviolet absorbance with an intense maximum in the range 220 nm and a weak one at 280 nm, depending on the polarity of their solvent (hexane 215 nm, methanol 221 nm, water 224 nm) (Fig. 5.5). This characteristic has been extensively exploited for the characterization of 4-hydroxynonenal produced during the peroxidation of hepatic tissues. Table 5.1 lists the molar absorption co-efficients for free hydroxyalkenals (average $\varepsilon = 10\,000$–$14\,000$ M$^{-1}\cdot$ cm^{-1} (Esterbauer and Weger, 1967).

Trace amounts of aldehydes, particularly acetaldehyde, propanal

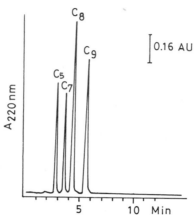

Fig. 5.6. HPLC of reference aldehydes. Reversed-phase column, Lichrosorb 5RP-18 (9.6 mm × 20 cm): 50°C: mobile phase acetonitrile/water (1:1, v/v): flow rate 0.92 ml/min: 222 nm detection. C_5 = hydroxypentenal; C_7 = hydroxyheptenal; C_8 = hydroxyoctenal; C_9 = hydroxynonenal. (Esterbauer, 1982.)

and butanal and to a lesser degree hexanal occur even in purified HPLC-grade solvents and pure water (Fig. 5.6). Because the levels of carbonyls produced during lipid peroxidation in vivo seldom exceed those found in the solvents used in derivatization and extraction, aldehydic impurities must be removed prior to analysis.

Preparation of carbonyl-free solvents
Carbonyl-free solvents can be prepared on Celite 545 columns impregnated with concentrated sulphuric acid according to the method of Hornstein and Crowe (1962).

REAGENTS
Celite 545 (BDH Chemicals Ltd.)
Sodium sulphate (anhydrous)
Sulphuric acid

PROCEDURE
A slurry containing 60 ml concentrated sulphuric acid and 100 g Ce-

lite 545 is stirred until homogeneous. A small volume of the solvent to be purified is poured into a 2.5 × 80 cm glass column equipped with a coarse fitted disc and stop-cock. Anhydrous granular sodium sulphate is then added to a depth of 10 cm and the slurried Celite 545 added to a depth of 30 cm. The column is finished with about 7–8 cm of cystalline sodium sulphate and the impure solvents poured through at a rate of 3–5 ml/min. It is recommended that chloroform is washed with water and dried over anhydrous sodium sulphate before preparation. Dichloromethane may become pink; this can be decolourized by shaking with activated alumina.

5.3.2.3. Methods for the characterization of hydroxyalkenals produced during lipid peroxidation

Free hydroxyalkenals are volatile and rather unstable so that considerable amounts can be lost during analysis through evaporation of solvents and elution from TLC plates. Determination of the free aldehyde has rarely been used for quantitative measurements, but the method is useful for qualitative characterization of the unsaturated carbonyls produced, during oxidative stress in vitro.

Extraction

Ethyl acetate has been used in the extraction of free aldehydic products of lipid peroxidation from liver homogenates and microsomal fractions challenged by oxidative stress in vitro (Benedetti et al., 1979; Esterbauer, 1982; Esterbauer et al., 1982). The incubation mixture is collected and extracted three times with an equal volume of ethyl acetate. The upper phases are collected and washed several times with an equal volume of water, and the surplus water is removed by freezing at −20°C. The ice crystals are then filtered on sintered glass and the solvent evaporated in a rotary evaporator under reduced pressure at 20°C.

Chloroform extraction has also been found to yield an almost complete recovery of hydroxyalkenals from peroxidized liver homogenates (Benedetti et al., 1986).

Thin-layer chromatography

Lipid residues are re-dissolved in ethylacetate and applied to a 20×20 cm plate coated with a 0.5 mm layer of Anasyl H (Analabs Inc., New Haven, CT, U.S.A.), which is then developed using a solvent system of n-heptane/ethyl acetate/acetic acid (72:25:1, v/v/v). Free hydroxyalkenals are detected as a yellow band (R_F value 0.33) after spraying the edge of the chromatoplate with a solution of N,N-dimethyl-p-phenylenediamine. The area corresponding to this band, in a part of the plate not exposed to the reagent is scraped off and the free carbonyls eluted first with a small volume of the developer, secondly with a solution of 0.5% (w/v) acetic acid in ethyl acetate and finally with n-heptane/ethylacetate/acetic acid (950:50:5, v/v/v). The solvents are subsequently pooled and evaporated under reduced pressure at 20°C.

High-performance liquid chromatography

Free hydroxyalkenals can be resolved in order of increasing chain length by HPLC using reversed-phase chromatography and a mobile phase of acetonitrile ('S' grade, Rathburns Chemicals, Scotland) and water.

HPLC CONDITIONS

Column	:	Lichrosorb RP 18 (5 μM) or Zorbax ODS (5 μm)
Column dimensions	:	20 cm × 4.6 mm, Lichrosorb RP 18 25 cm × 4.5 mm, Zorbax ODS
Mobile phase	:	Acetonitrile/water (1:1, v/v)
Flow rate	:	0.92 ml/min
Temperature	:	50°C
Detection	:	UV 220 nm
Sample injection	:	50 μl $\rightleftharpoons$ 5 mg liver
Analysis times	:	Lichrosorb RP 18 < 15 min Zorbax ODS < 10 min
Sensitivity	:	0.5 nmol 4-HNE at 0.01 aufs

(See Figs. 5.6 and 5.7.)

STANDARDIZATION

4-Hydroxyalkenals are not commercially available but can be synthesized (Esterbauer and Weger, 1967; Erickson, 1974; Benedetti et al., 1982; Gree et al., 1986). Aqueous solutions of hydroxyalkenals are stable for about 1 week at 4°C, but decompose within 24 h at room-temperature. They are stable if stored at −20°C in chloroform.

5.3.2.4. Determination of free 4-hydroxy-2,3-trans-alkenals by HPLC Esterbauer (1982) has developed a procedure for the 'qualitative detection and quantitative measurement of steady-state concentrations' of free hydroxyalkenals (specifically HNE) in tissues, tissue extracts and lipid containing foodstuffs. Their method utilizes UV-detection of the free aldehyde at its 220 nm UV-absorption maximum and peak identification was confirmed by mass spectrometry. An effective purification and concentration step is employed using dichloromethane to extract hydroxyalkenals from samples 'trapped' on 'Extrelut' columns. The samples are subsequently purified by solid-phase extraction on octadecyl-bonded silica (ODS) disposable cartridges and then analysed by HPLC.

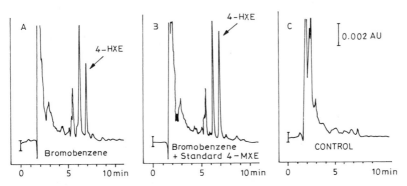

Fig. 5.7. HPLC chromatography of 4-hydroxynonenal (HXE) as free aldehyde in livers of bromobenzene-poisoned mice. (**A**) bromobenzene-treated mice; (**B**) as A with 5 pmol standard 4-NE added; and (**C**) control mice. Zorbax ™ODS (4.6 mm × 25 cm): 50°C: mobile phase acetonitrile/water (1:1,v/v): flow rate 1 ml/min 222 nm detection. (Esterbauer et al., 1982.)

The major advantages of the method are that the samples do not need to be derivatized and tedious semi-preparative chromatography of the dinitrophenylhydrazones is unnecessary (Fig. 5.7). The method is very precise with co-efficients of variation ranging from 1.4 to 3.5% for free 4-hydroxy-2,3-*trans*-nonenal (HNE). The recovery of 4-hydroxynonenal from microsomes, which was assessed by addition of known amounts of [^{14}C]HNE to microsome suspensions and measurement of the radioactivity in extracts ready for HPLC is poor, ranging from 45–73% for normal and peroxidized tissues. This is due to an incomplete extraction of hydroxyalkenals from membrane phospholipids with which they may form covalent linkages. HNE for example is extremely reactive towards sulphydryl groups of proteins and glutathione with which it forms adducts (Benedetti et al., 1979; Benedetti et al., 1980); HNE also reacts with cysteine to form thiazolidine derivatives (Esterbauer et al., 1982), with ε-amino groups of some lysine residues in proteins (Benedetti et al., 1984) and with phosphatidylethanolamine (Esterbauer et al., 1982).

Procedure
MATERIALS AND REAGENTS
Extrelut columns for a 20 ml sample volume, containing Kieselguhr as stationary phase (Merck, Darmstadt, F.R.G.).

Octadecyl silice gel columns (3 ml size) (J.T. Baker, Phillipsburg, NJ, U.S.A.) or Sep-Pak, ODS disposable columns (Waters Associates).

2,6-Di-*tert*-butyl-*p*-cresol (BHT) in ethanol (10 mg/ml) (Sigma Chemicals)

Desferal (desferrioxamine; 10 mg/ml in distilled water) (Ciba-Geigy, Horsham, U.K.)

Methanol
Acetonitrile ⎫
Hexane   ⎬ HPLC grade, Rathburns Chemicals, Scotland
Dichloromethane ⎭

Acetate buffer 0.1 M (pH 3.0)

PREPARATION OF SUBCELLULAR ORGANELLES AND CELLS FOR SOLID-
PHASE EXTRACTION

Samples are prepared for solid-phase extraction by the addition of 20
μl ethanolic 2,6-di-*tert*-butyl-*p*-cresol (BHT, 10 mg/ml) and 20 μl of
aqueous desferrioxamine (10 mg/ml) to 20 ml of a suspension of subcellular organelles ($\sim$1 mg protein/ml) to 20 ml. The inclusion of the
antioxidant BHT and the iron chelator desferrioxamine prevents lipid
peroxidation. The suspension is poured onto the Extrelut column and
'loaded' by applying a vacuum at the column outlet; lipids are eluted
with 40 ml dichloromethane into a flask containing 2 ml of 0.1 M acetate buffer (pH 3.0). Interaction of hydroxyalkenals with phospholipids in the dichloromethane, and loss by evaporation under vacuum,
are thus prevented. Dichloromethane is removed to half the original
volume under vacuum at 20°C in a rotary evaporator. The dichloromethane extract contains the carbonyls, together with phospholipids,
fatty acids, triglycerides, cholesterol and cholesteryl esters. If whole
tissues are to be analysed, they are cut into small pieces and 10 ml
water containing 0.1% BHT (w/v) and 1% (w/v) desferal is added. The
tissues are homogenized for 2 min with an Ultraturrax blender and
centrifuged at $3000 \times g$ for 10 min. The supernatant is decanted and
the residue re-extracted with 10 ml water containing BHT and desferal. The combined supernatants are then purified by extraction as
described above.

Oil samples (1 g) can be treated with 10 ml distilled water as described for whole tissues and the mixture shaken for 20 min. The
aqueous extracts are then treated as described for whole tissues.

Purification of the dichloromethane extract
The ODS extraction column is preconditioned with 3 ml methanol
and 3 ml water. The eluants from the Extralut columns are then applied to the ODS columns which are eluted with either 2 ml (for organelles) or 15 ml (for oils, tissues) hexane to remove non-polar materials which subsequently otherwise interfere during HPLC analyses.
Hydroxyalkenals are subsequently extracted with 2 ml methanol/
water (80:20) into a 2 ml volumetric flask and residual hexane in the

eluant is removed under nitrogen at 20°C and the sample is brought
to 2 ml by addition of methanol/water (80:20). As an alternative, Sep-
Pak, ODS cartridges which can be attached to the Luer fitting of glass
syringes can be used without further apparatus.

HPLC CONDITIONS
Column   : Spherisorb S5 ODS 2
Column dimen-   : 25 cm × 0.48 cm
sions
Mobile phase   : *either* Acetonitrile/water (40:60, v/v)
                        or Methanol/water (65:34, v/v)
Flow rate   : 1.0 ml/min
Temperature   : ambient
Detection   : UV 220 nm
Sample injection   : 20 μl
Analysis times   : about 6 min (HNE)
Sensitivity   : approx. 0.05 μM

STANDARDIZATION
The molar absorption coefficient of hydroxyalkenals in water and ethanol or methanol is 13 750 at 223 nm and 13 100 at 221 nm, respectively (Esterbauer, 1982). Calibration curves of peak height versus concentration are linear in the range 0.1–500 μM.

5.3.2.5. Analysis of aldehydic products of lipid peroxidation as
2,4-dinitrophenylhydrazone derivatives
In order to enhance the detectability of saturated aliphatic carbonyls

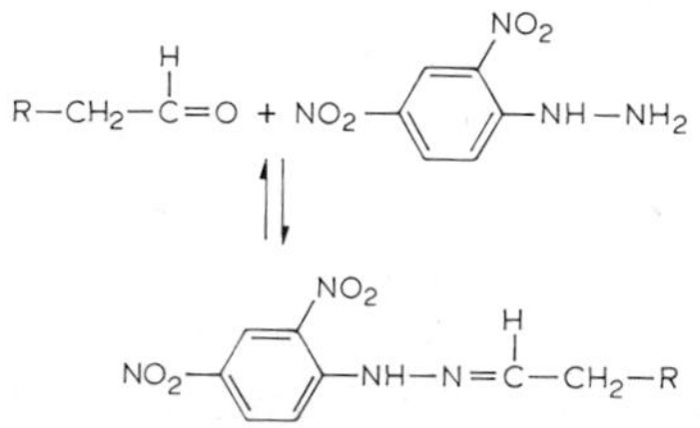

Fig. 5.8. Reaction of carbonyls with 2,4-dinitrophenylhydrazine.

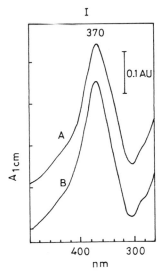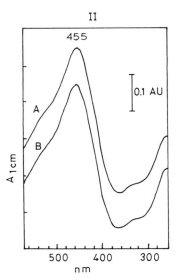

Fig. 5.9. Ultraviolet-visible absorption spectrum of the 2,4-dinitrophenylhydrazones of **(A)** biogenic 4-hydroxynonenal; **(B)** synthetic 4-hydroxynonenal; I: in CHCl$_3$; II: in 0.1 M ethanolic KOH. (Esterbauer, 1982.)

and reduce the volatility of hydroxyalkenals, precolumn derivatization has been used.

The most widely used reagent is 2,4-dinitrophenylhydrazine; carbonyls react with 2,4-dinitrophenylhydrazine according to the scheme in Fig. 5.8.

The reaction is quantitative and at gram quantities, the hydrazone precipitates immediately due to its low solubility in the aqueous phase. The removal of the derivative from the aqueous phase shifts the equilibrium toward the formation of more derivate (Esterbauer, 1982). The dinitrophenylhydrazones absorb strongly in the region 360–390 nm. The spectrum of the 2-alkenals is shifted to 455 nm in ethanolic KOH (Fig. 5.9).

The latter characteristic can be exploited for identification of putative 2-alkenals in biological samples at levels too low for more specific mass-spectrophotometric analyses. The ε values for carbonyl dinitro

TABLE 5.2.

Class	ε (litre $\cdot$ mol^{-1} $\cdot$ cm^{-1})	λ_{Max} (nm)
4-Hydroxyalkenals	28 000	370
Osazones	44 000	430
Alkanones	225 000	362
n-Alkanals	21 600	360
Alk-2-enals	28 000	378
Alka-2,4-dienals	37 000	392

A scheme for the analysis of complex aldehyde mixtures is as follows (after Esterbauer, 1982) (Scheme 5.2).

phenylhydrazones are given in Table 5.2 (from Benedetti et al., 1984).

5.3.2.5.1. Derivatization of aldehydes using 2,4-dinitrophenylhydrazine The recommended method is that of Benedetti et al. (1984) as subsequently modified by Poli et al. (1985). Not all the free carbonyls produced are quantitatively extractable in chloroform or dichloromethane, and the 2,4-dinitrophenylhydrazones are best prepared before extraction using tissue homogenates and incubation mixtures directly (Benedetti et al., 1986).

PREPARATION OF 2,4-DINITROPHENYLHYDRAZINE REAGENT
A suspension of 2,4-dinitrophenylhydrazine (0.05%, w/v) in 1 M HCl is warmed for 1 h at 50°C, stirred, cooled to room temperature and extracted with n-hexane (2 × 50 ml) in order to extract trace carbonyl impurities. The reagent is flushed with nitrogen, stoppered and stored in the dark at ambient temperature for not more than 24 h.

DERIVATIZATION
The dinitrophenylhydrazones are prepared by mixing equal volumes of sample and 2,4-dinitrophenylhydrazine reagent. After 2 h the reaction mixtures are extracted twice with an equal volume of dichloromethane, centrifuged at 2000 × g for 10 min, and the combined solvent evaporated under nitrogen at reduced pressure at 35°C in a rotary evaporator.

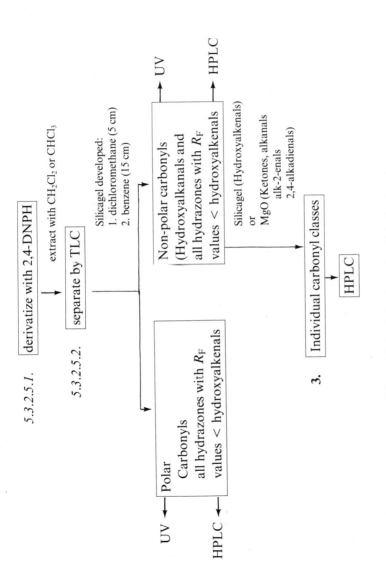

Scheme 5.2 Analysis of complex aldehyde mixtures.

5.3.2.5.2. Separation of aldehyde dinitrophenylhydrazines by TLC and column chromatography   Excess dinitrophenylhydrazine reagent and the hydrazones of acetone, formaldehyde and acetaldehyde, which occur as contaminants, can be removed prior to preparative TLC by column chromatography. Silicic acid (Merck) columns (3×2.1 cm) may be used with chloroform to elute the aldehyde dinitrophenylhydrazones.

PREPARATIVE TLC
Carbonyl derivatives can be separated into three groups by TLC on silica-gel 60 plates (20×20 cm) using dichloromethane (to 5 cm) and subsequently benzene (to 15 cm) as developer. Better separations may be achieved by further development in benzene but this can only be determined by trial. The three groups of carbonyls are resolved into three zones (Benedetti et al., 1986).

Zone 1: from the origin to R_F 0.12; considered to be the hydrazones of the hydroxyalkenals polar carbonyls;

Zone 2: inclusive of the hydrazones of the hydroxyalkenals and all components less mobile than unreacted 2,4-dinitrophenylhydrazine;

Zone 3: non-polar carbonyls including the alkanals, and alk-2-enals. These can be further resolved into two fractions comprising osazones and a mixture of alkanals and alkenals using silica-gel 60 and chloroform/hexane (7:3, v/v) as developer. In all cases, the hydrazones corresponding to standards run concomitantly are scraped off and eluted in chloroform/methanol (9:1, v/v).

PREPARATIVE HPLC
Normal-phase HPLC may be used for separation of some aldehyde classes prior to quantitative reversed-phase HPLC. It is possible to separate 4-hydroxyalkenals by normal-phase HPLC as described below using 50% water-saturated dichloromethane as a mobile phase. This is prepared by mixing equal volumes of dry dichloromethane and a solution of dichloromethane stored under a layer of water (100%

water-saturated solution). To remove water droplets that remain with the 100% saturated solution, it is decanted twice before the addition of dry solvent.

HPLC CONDITIONS

Column	:	Lichrosorb S:60
Column dimensions	:	15 cm × 0.48 cm i.d.
Mobile phase	:	50% water saturated dichloromethane
Flow rate	:	1.5 ml/min
Temperature	:	ambient
Detection	:	378 nm
Sample injection	:	100/200 μl in mobile phase (100 μl loop)
Retention time	:	10–13 min

The disadvantage of normal phase HPLC for the separation of carbonyl dinitrophenylhydrazones is the difficulty in maintaining reproducible retention times due to absorption of water from the solvent and interaction with the active groups on the column (Fig. 5.10). To overcome this problem, a cyano-propyl bonded phase (Exsil 100, 5 μm; Chromtech, U.K.) column may be used for isolation of hydroxyalkenals and a mixture of alkanals and alk-2-enals from rat liver microsomes. The less polar carbonyls can be separated under isocratic conditions as follows.

HPLC CONDITIONS

Column	:	Exsil 100CN 5 μm
Column dimensions	:	25 cm × 0.46 cm
Mobile phase	:	Isocratic for 10 min
		Hexane: Isopropanol (95:5, v/v) rising to 100% isopropanol in 8 min
Flow rate	:	1.0 ml/min
Temperature	:	ambient
Detection	:	378 nm

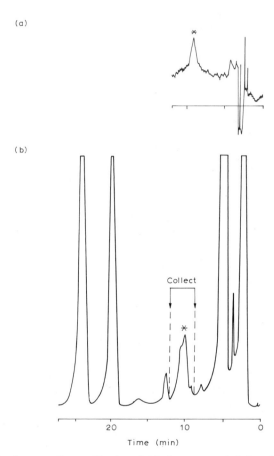

Fig. 5.10. Semi-preparative purification of 4-hydroxynonenal dinitrophenylhydrazone in microsomal fraction from 0.5 g rat liver. Reversed-phase column, Lichrosorb Si60 (51 m); solvent: dichloromethane (50% in water; flowrate: 1.5 ml/min; detection: 340 nm; injection: 100–200 μl; **(a)** Standard; **(b)** Liver microsomal fraction. (Benedetti et al., 1986.)

Sample injection   :   100/200 μl in mobile phase (100 μl loop)

Retention time   :   Alkanals + alk-2-enals 8–14′
Hydroxyalkenals   18–20′

The disadvantage of both methods is that the carbonylhydrazones cannot be monitored directly because of masking by interfering peaks. To overcome this problem, fractions are collected not only by reference to pure standards but by reference also to the chromatographic pattern given by contaminants in a small volume of sample to which pure standards had been added. In this procedure, the 'finger-print' pattern of the chromatographic peaks is compared to that given by the sample alone.

QUANTITATIVE ANALYSIS

After the separation of carbonyl dinitrophenylhydrazones into classes or into polar and non-polar fractions, the samples can be re-dissolved in a small volume of methanol and the amount present calculated approximately from the absorbance maxima at 365–370 nm using an average molar absorption coefficient of 26 000 (Esterbauer, 1982). Alternatively, the amount of carbonyls within individual classes can be calculated from the molar absorptivities shown in Table 5.1.

REVERSED-PHASE HPLC

The complex mixtures of hydroxyalkenals separated by TLC on silica gel, and the other individual carbonyl classes separated by TLC on MgO can be determined by reversed-phase HPLC.

HPLC CONDITIONS

Column	:	Zorbac ODS (5 μm)
Column dimensions	:	25 cm × 0.46 cm (i.d.)
Mobile phase	:	Methanol/water (92.5:7.5, v/v)
Flow rate	:	1.2 ml/min
Temperature	:	ambient
Detection wavelength	:	378 nm
Sample injection	:	in mobile phase
Retention times	:	4–16 min

Retention times can be lengthened by increasing the concentration of water in the mobile phase (Fig. 5.11).

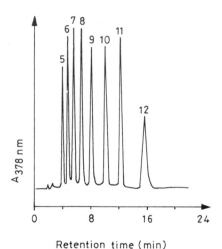

Retention time (min)

Fig. 5.11. HPLC separation of reference *n*-alkanal-diphenylhydrazones; the numbers indicate the chain length of the aldehyde thus: 5: pentanal; 6: hexanal etc. Zorbax ODS column (4.6 × 250 mm); methanol/water 92.5:7.5, 1.2 ml/min, 378 nm detection. (Benedetti et al., 1986.)

ALKANALS

Individual alkanal dinitrophenylhydrazones are eluted in order of decreasing polarity.

ALK-2-ENALS

Individual alk-2-enal dinitrophenylhydrazones can be resolved under the conditions described for individual alkanals using a mobile phase of methanol/water (9:1, v/v). The carbonyls are eluted in order of decreasing polarity between 4 and 12 min after injection (Fig. 5.12).

4-HYDROXYALKENALS

These can also be separated as described for individual alkanals using either a column packed with Lichrosorb 5-RP18 (20 × 0.46 cm i.d.) and a mobile phase of methanol/water (4:1, v/v) or Spherisorb 5-ODS (25 × 0.46 cm i.d.) with a similar mobile phase in the proportions 3:9 (v/v). In both cases, 4-hydroxynonenal is resolved within 16 min (Fig. 5.13).

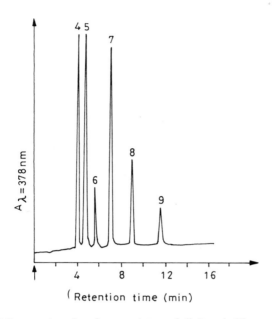

Fig. 5.12. HPLC separation of a reference mixture of alk-2-enals. The numbers indicate the chain length of the aldehyde, thus; 4: but-2-enal; 5: pent-2-enal etc. (Benedetti et al., 1986.)

5.3.3. Assay of blood and tissue 2,4-dinitrophenylhydrazones

Lynch et al. (1983) described a method for the assay of acetaldehyde in blood. Dinitrophenylating reagent is prepared freshly by dissolving 3 mg of 2,4-dinitrophenol in a solution of 60 ml methanol in 20 ml double-distilled, deionized water with 0.5 ml sulphuric acid. Either 1 ml blood or 1 g tissue is mixed with 10 ml reagent containing 50 μl methanolic 2,4-dinitrophenylhydrazine[14C]formaldehyde (prepared by adding 1 ml dinitrophenylating reagent to [14C]formaldehyde (10–20 mCi/mmol; Amersham International). The crystalline adduct is washed with 2 ml 2 M H_2SO_4, dried and desiccated. After vortex mixing or homogenization in a Polytron homogenizer, the samples are centrifuged at $1700 \times g$ for 15 min and the supernatant removed. Ex-

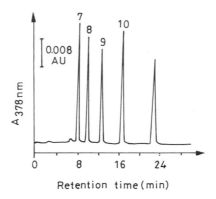

Fig. 5.13. HPLC separation of a reference mixture of 4-hydroxy-2-alkenal dinitrophenyl hydrazones. The numbers indicate the chain length of the aldehyde thus; 7: 4-hydroxy-2-heptenal; 8: 4-hydroxy-2-octenal etc. Lichrosorb 5-RP-18, (4.6 nm × 20 cm); methanol/water 80:20 (v/v); 1.0 ml/min; detection 378 nm. (Benedetti et al., 1986.)

cess dinitrophenylhydrazine reagent is quenched by addition of 0.5 ml 2.7 M aqueous formaldehyde and the adducts extracted into 2 × 5 ml portions of chloroform. The solvent is then washed with 2 M HCl (2 × 10 ml) and water (2 × 10 ml) and the chloroform extract purified by aluminium hydroxide chromatography (3 × 0.5 cm in a Pasteur pipette) and evaporated under nitrogen. Finally, the samples are re-dissolved in 0.5 ml methanol for HPLC analysis and 0.2 ml aliquots taken for scintillation counting (as an indicator of losses during analysis). Acetaldehyde is detected at 356 nm after separation on ODS Hypersil (5 μm, 25 × 0.46 mm i.d.) with a mobile phase of acetonitrile/water (1:1, v/v) at a flow rate of 1 ml/min. The limit of detection is 0.1 μM.

5.3.4. Analysis of aldehydic products of lipid peroxidation as fluorescent decahydroacridine derivatives by HPLC

5,5-Dimethyl-1,3-cyclohexanedione (dimedone) and 1,3-cyclohexanedione (CHD) are highly specific and extremely sensitive reagents which condense with aldehydes to produce fluorescent compounds. The maximum excitation and emission wavelengths are very similar for most aldehydes and are in the range 383–393 nm and 455–462 nm, respectively for straight-chain aliphatic aldehydes. Aldehydes exhibiting branching or unsaturation at the α position do not fluoresce with the same intensity. 1,3-Cyclohexanediones have been used for trace analysis of aldehydes in environmental, industrial and food samples (Mopper et al., 1983; Suzuki, 1985), rat liver microsomes (Yoshino et al., 1986a; Yoshino et al., 1986b; Tomita et al., 1987), muscle homogenates and homogenates of brain stem (Petonen et al., 1984). The decahydroacridine derivatives can be resolved by reversed-phase HPLC. Both assay selectivity and detection sensitivity are significantly better than can be achieved by measurement of dinitrophenyl-hydrazones.

Derivatization procedure using 1,3-cyclohexanedione

REAGENTS
1,3-Cyclohexanedione (Aldrich Chemical Co.)
Ammonium sulphate
Glacial acetic acid

The derivatizing reagent is prepared by dissolving 0.25 g CHD and 10 g ammonium sulphate in 50 ml of Milipore-filtered water. After the addition of 5 ml of acetic acid, the reagent is brought to 100 ml volume with water and the pH adjusted to 5.0. This reagent is stable for several weeks at room temperature when protected from the light.

PROCEDURE

Methanolic solutions of pure aldehydes (0.5 ml) are mixed with an equal volume of water and 2 vol of reagent. Plasma or tissue homogenates are mixed with an equal volume of methanol, vortex mixed for 30 s, centrifuged at $850 \times g$ for 10 min. The supernatant (0.5 ml) is mixed with 1 ml reagent, and the samples are heated for 1 h at 60°C. The derivatives can be extracted quantitatively from the aqueous reaction mixture utilizing Sep-Pak C18 disposable solid-phase extraction cartridges (Waters Associates) which have been pre-conditioned with 2 ml of methanol and 5 ml of water. The aqueous mixture is poured into the cartridge, washed with water and the derivatives eluted with 2.0 ml of methanol. The derivatives are stable indefinitely in the organic phase.

HPLC CONDITIONS

Column	:	ODS Hypersil (5 μm)
Column dimensions	:	15 cm × 0.46 cm
Mobile phase	:	Linear gradient program
		50% Methanol/water (30:70, v/v)
		50% Tetrahydrofuran/water (26:74, v/v) rising to 100% THF-water (26:74, v/v) after 9 min
Flow rate	:	1.0 ml/min
Temperature	:	ambient
Detection	:	fluorescence: excitation at 380 nm emission at 445 nm
Sample injection	:	10–20 μl in THF/H$_2$O (26:74, v/v)
Limits of detection	:	10 pg for aliphatic aldehydes. Calibration curves for aliphatic alkanals are linear in the range 0–10 ng

Yoshino et al. (1986a) resolved a number of aldehyde cyclohexane dione combinations using an ERC-ODS-1262 column (5 μm, 10 × 0.6 cm; Emma Optical Works Ltd., Tokyo, Japan) and a gradient elution profile with methanol/water and tetrahydrofuran/water mixtures as follows;

0–18 min   MeOH/H$_2$O (30:70)
18–32 min   THF/H$_2$O (26:74)
32–42 min   THF/H$_2$O (40:60)
42–50 min   100% THF

Mopper et al. (1983) separated aldehydes after derivatization with 5,5-dimethyl-1,3-cyclohexanedione (dimedone) by reversed-phase HPLC on either Altex Ultrasphere ODS (5 μm, 25 × 0.46 cm) or on Nucleosil ODS (5 μm, 20 × 0.46 cm). C_1–C_7 aldehydes were separated by isocratic elution using acetonitrile/water (60:40, v/v). The derivatives can also be resolved by normal-phase HPLC using porous silica (Nucleosil, 7 μm, 20 × 0.46 cm) with a mobile-phase of hexane/isopropanol (99:1, v/v) eluted at 1.0 ml/min. The reagent is prepared by dissolving 60 g of ammonium acetate and 2.1 g of dimedone (in isopropanol 0.12 g/ml) in 100 ml of water. The solution is stored in an amber bottle and left to stand overnight; thereafter the reagent is stable for several weeks at room temperature. To eliminate aldehyde contamination, the reagent is heated at 100°C for 30 min in a sealed vessel and then extracted with 2 × 20 ml carbonyl-free dichloromethane. Derivatives are prepared by addition of 2 ml of aqueous sample to 1 ml of reagent and 0.25 ml of 9 M sulphuric acid (made carbonyl free by refluxing for 2 h) and the pH adjusted to 5.0. The reaction tube is sealed and heated to 100°C in a water bath for 20 min. Thereafter, the derivatives are extracted with 0.6 ml of carbonyl-free dichloromethane or by passage through a C18 Sep-Pak as described for the CHD reagent.

5.4. *Alkane and alkene measurement in exhaled breath and isolated organs or organelles by gas liquid chromatography (GLC)*

The use of GLC for the measurement of hydrocarbon gases evolved following lipid peroxidation was pioneered by Riely et al. (1974). The most common hydrocarbons that have been measured are ethane, which is evolved during the oxidation of $n-3$ fatty acids and pentane which is a decomposition product of $2n-6$ fatty acid hydroperoxides. Other hydrocarbons are also formed in smaller amounts including ethylene from linoleic acid, octane from oleic acid, heptene and hexane from vaccenic acid and pentene and butane from myristoleic acid (Tappel, 1980).

$$H_3C\text{-}CH_2\text{-}\underset{\underset{O^{\bullet}}{|}}{CH}\text{-}R \longrightarrow H_3C\text{-}\overset{\bullet}{C}H_2 + \underset{\underset{H}{|}}{\overset{\overset{O}{\parallel}}{C}}\text{-}R$$

$$H_3C\text{-}\overset{\bullet}{C}H_2 + RH \longrightarrow H_3C\text{-}CH_3 + R^{\bullet} \qquad \text{(From Evans et al., 1969)}$$

Tappel has suggested that pentane provides the best index of peroxidation in vivo because of the quantitative prevalence of linoleic acid in biological tissues (Dillard et al., 1977; Tappel, 1984). Alipathic hydrocarbons can be hydroxylated by microsomal monooxygenases (Frommer et al., 1970), but Lawrence and Cohen (1984) found that, whereas ethylene and C_3–C_5 straight-chain hydrocarbons were rapidly eliminated from expired air by rats and mice in enclosed chambers, ethane was not. This suggests that ethane is a better index of lipid peroxidation in vivo (Riely et al., 1974; Burk and Ludden, 1984). The main advantages in the measurement of exhaled hydrocarbon gases as an index of lipid peroxidation in vivo include:

(a) The method is non-invasive and utilizes whole animals (Lawrence and Cohen, 1984) or human subjects (Lemoyne et al., 1987; Van Gossum et al., 1988) which can be monitored sequentially over days or months;

(b) the method can be adapted for use with isolated organs and cells (Muller and Sies, 1984) or subcellular organelles (Kappus and Mulliawan, 1988; Wendel, 1987); derived from animals that have been used for whole animal studies.

(c) there is no likelihood of the autoxidation of samples during their preparation.

Several factors beside the rate of peroxidation contribute to the rate of alkane exhalation:

(a) Bacterial peroxidation of dietary PUFAs in the gut of animals can yield increased rates of hydrocarbon formation;

(b) inhaled air is frequently contaminated with high background levels of hydrocarbons:

(c) some hydrocarbons can be metabolized in vivo.

Recent studies have shown that alkane exhalation does not correlate with other indices of lipid peroxidation such as diene-conjugation and free and bound MDA in well-oxygenated organelles and organs such as the lung (Reiler and Burk, 1984; Kostrucha and Kappus, 1986; Van Gossum et al., 1988). This may be because the oxygen concentration at the site of alkane formation affects the molar yield of hydrocarbons; a rise in oxygen concentration from 5–100% causes a 100-fold fall in molar yield of alkanes (Reiter and Burk, 1987). It has been suggested that oxygen reacts with alkyl radicals more rapidly than hydrogen abstraction can occur leading to oxygen-containing products other than ethane and pentane (Cohen, 1982).

$$H_3C\text{-}\overset{\bullet}{C}H_2 + O_2 \rightarrow H_3C\text{-}CH_2OO^{\bullet}$$

Nevertheless, high rates of alkane exhalation are usually indicative of lipid peroxidation in vivo but small increases should be interpreted with caution, especially when using the method as a quantitative measure of lipid peroxidation in vivo.

5.4.1. Exhalation of hydrocarbons by animals

Experimenters with animals have adopted two practical approaches. The alkanes are excreted in small concentrations (nanomoles/litre of expired air) and it is usual to concentrate the gases before analysis by trapping the alkanes by concentration of a large volume of gas by condensation on cold precolumns. In a typical experiment the head of a small laboratory animal is enclosed within a gas-tight collar and the animal allowed to breathe in a continuous air flow (Wendel, 1987). This 'open system' prevents contamination of the effluent air with fecal, intestinal and other gases excreted by microorganisms. The method also reduces the metabolism of exhaled hydrocarbons produced by the animal other than that achieved in vivo during the 'first pass' from the site of origin through the lungs. Hydrocarbons in the effluent air are absorbed on an activated surface such as alumina. This has been achieved using a sample loop (0.5 ml volume) immersed in a cold 'slush' of ethanol/liquid nitrogen at $-130°C$. The loop is connected directly to a gas chromatograph and hence the concentrated sample can be directly injected for analysis.

Lawrence and Cohen (1984) found that ethane was not well absorbed by alumina. They described a method for concentrating ethane, ethylene and longer chain hydrocarbons contained in up to 50 ml of air. They used a closed system in which the air, exhaled by mice allowed to breathe hydrocarbon scrubbed air was filtered through sulphuric acid and potassium hydroxide to trap ammonia and carbondioxide, and condensed on a 'cold finger' immersed in solid $CO_2/$

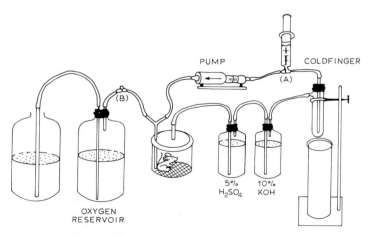

Fig. 5.14. Breath collection chamber and air circulation system for measurement of exhaled volatile hydrocarbons (Lawrence and Cohen, 1984.)

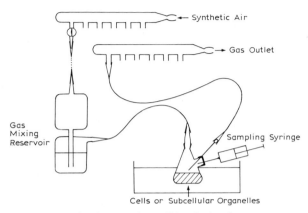

Fig. 5.15. System used with microsomal or cell incubation for measurement of evolved volatile hydrocarbons. (Muller and Sies, 1984.)

propan-2-ol. Samples of effluent air were collected using 50 ml gas-tight syringes and the hydrocarbon content concentrated using a Chemical Data Systems (Oxford, PA, U.S.A.) model 310 concentrator. This contained both alumina on which pentane and long-chain hydrocarbons were absorbed and activated charcoal (type SK-4, 60/80 mesh) on which ethane and ethylene were adsorbed. The hydrocarbons were subsequently desorbed by heating at or above 220°C (Fig. 5.14)

5.4.2. Methods utilizing isolated, perfused whole organs and cells or cell organelles

Muller and Sies (1984) constructed a simple apparatus consisting of a glass manifold through which alkane-free air passed into a glass flask filled with water and then into a 20 or 60 ml gas-tight Erlenmeyer flask with side arm (Fig. 5.15).

Cells and cell-organelles were incubated at 37°C in the Erlenmeyer flask, and 8 ml aliquots withdrawn from the gas-phase at intervals using Hamilton gas-tight syringes. Aliquots were then injected into the sampling loop of a gas-chromatograph.

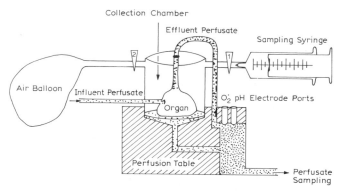

Fig. 5.16. System used for measurement of volatile hydrocarbons produced by isolated perfused liver (non-recirculating perfusion). (Muller and Sies, 1984.)

An apparatus suitable for use with isolated perfused organs is shown in Fig. 5.16.

5.4.3. Human studies

An apparatus suitable for human breath collection was described by Lemoyne et al. (1987). Subjects were allowed to breathe for 4 min through a mouth piece connected to a Rudolph Valve from a Tedlar bag (Analygas Systems Ltd., Scarborough, Ontario, Canada) containing hydrocarbon-free air. Atmospheric air was flushed from the lungs and an aliquot of exhaled air collected during the succeeding 2 min while the hydrocarbon-free air was inspired. Hydrocarbons were concentrated by a loop-concentrator similar to that previously described and then injected into the GLC.

A simple method for the analysis of small volumes of gas from single-breath samples from humans was described by Zarling and Clapper (1987). Total breath samples were collected into a gas-tight bag and 50 ml aliquots withdrawn into polyethylene/polypropylene syringes. Alveolar breath samples collected by use of a Haldane-Priestly tube were also collected in this way. Gas samples were injected directly into the GLC via a gas-sampling valve.

5.4.4. Chromatographic conditions

Many different instruments have been used for the measurement of hydrocarbons in exhaled air. They include the Varian Model 6000 (Varian Instrument, Sunnyvale, CA, U.S.A.) (Zarling and Clapper, 1987), Shimadzu 6-AM (Shimadzu, Seiggkuska Ltd., Kyoto, Japan) (Lemoyne et al., 1987) and Hewlett-Packard Model 5750 (Lawrence and Cohen, 1984). Gas samples are injected either directly into a sampling loop (10 ml volume) (Zarling and Clapper, 1987) or via stainless steel loops (0.2 cm i.d. × 23.4 cm) packed with adsorbents such as activated alumina (80–100 mesh) (Lemoyne et al., 1987) in order to concentrate the alkanes in the sample.

The choice of column and operating conditions varies with the degree of resolution required. Nitrogen is generally used as carrier gas which has been further purified by passage through a cartridge containing a desiccant (Drierite, WA Hammond Drierite Co., Xenia, OH, U.S.A.) and a carbon dioxide adsorber (Carbosorb AS, British Drug Houses, Toronto, Ontario, Canada) (Lemoyne et al., 1987). Isothermal determination of ethane and pentane may be accomplished at 50°C on a 3.2 mm × 2 m stainless-steel column packed with Porasil B (80/100 mesh, Applied Science Labs) (Lawrence and Cohen, 1984). Improved resolution of ethane from methane (produced by the gut flora of laboratory rodents) was achieved on a longer (4–6 m) column. Methane, ethane and ethylene can be further resolved using a 3.2 mm × 2 m stainless-steel column packed with Poropad N (60–80 mesh; Applied Science Labs.) operated at 60°C; pentane can be desorbed from this column by elevating the oven temperature to 140°C. Isothermal analyses have also been used to resolve ethane from exhaled air, using a Porasil C Column (80–100 mesh) operated at 50°C with a carrier gas flow of 40 ml/min and an air and hydrogen flow through the FID of 450 and 35 ml/min (Muller and Sies, 1984) and pentane can be desorbed at 90°C.

Concomitant measurement of alkanes in exhaled air may be achieved using a 2 metre Chromosorb 102 stainless-steel column (Varian Associates) with a carrier gas flow of 30 ml/min, an injector temperature of 150°C and a detector temperature of 225°C (Zarling and Clapper, 1987). A temperature program is used to minimize peak-broadening due to the injection of large sample volumes (10 ml). After injection, the column is operated for 1 min at 50°C and then the temperature increased to 100°C at a rate of 50°C/min. Finally, the temperature is allowed to rise to 190°C at a rate of 15°C/min. In a total run time of 15 min, it is possible to resolve ethane (retention time 2.67 min), propane (4.56 min), butane (7.74 min) and pentane (8.85 min). The method may be calibrated using standard gas samples (Alltech Associates, Deerfield, IL, U.S.A.) at concentrations of 0–13 nmol/litre of air. Pentane has been measured on a 3 mm × 2.4 m stainless-steel column packed with Porasil-D, operated isothermally at 60°C with a carrier gas flow of 20 ml/min, and a detector temperature of 200°C. The air and hydrogen gas rates are 800 ml/min and 30 ml/min (Lemoyne et al., 1987).

CHAPTER 6

Assay of antioxidant nutrients and antioxidant enzymes

6.1. Measurement of vitamin E

6.1.1. Introduction to methodology

The increasing interest in α-tocopherol as the principal lipid-soluble antioxidant in mammalian cells creates a need for rapid, reproducible and accurate methods for measurement of the vitamin in body fluids, cultured cells, whole tissues and, in some instances, in food products. Early methods identified a number of difficulties that arise in achieving this analytical objective and an understanding of the possible problems that may arise is fundamental to the successful application of the chosen analytical methods. The early methods were summarized in three categories (Bunnell, 1971) which cover the three stages of analysis that may still be found to be necessary with the availability of modern analytical techniques. These stages are:

(a) **Extraction.** α-Tocopherol is readily extracted from body fluids such as serum or plasma by simple organic solvents. Difficulties arise with tissues, which may require quite drastic homogenization or maceration; the maceration procedure itself carries risks in that surfaces may be continuously exposed to oxygen which may lead to oxidative degradation of the tocopherols. Even greater difficulties arise with the analysis of foodstuffs in which the tocopherols may be present in very low concentration in a matrix consisting of many different carbohydrate and protein components, as well as high concentrations of other lipids which may themselves be subject to oxidative degradation (e.g., unsaturated fats) which will cause destruction of tocopherols during the analytical procedure.

(b) Separation. The removal of other lipid components followed by the separation of the tocopherols one from the other present different problems. Alkaline saponification of the triglycerides has been used extensively to remove much of the interfering lipid. This technique is fraught with danger because of the inherent instability of the tocopherols in alkaline solution; the problem can be largely overcome by employing a high concentration of alkali, and by incorporating an antioxidant in the saponification mixture. Ascorbic acid and a range of synthetic antioxidants such as propyl gallate, butylated hydroxyanisole and butylated hydroxytoluene have been recommended but high recoveries of added tocopherols may only consistently be obtained with pyrogallol. The unsaponifiable fraction is extracted into a non-polar solvent such as hexane and the residual alkali and soaps removed by washing with water, care being taken to prevent emulsion-formation.

The principal constituents of the unsaponifiable fraction are, in addition to the tocopherols which may be present in relatively small amounts, vitamin A, cholesterol, ubiquinone and ubichromenol. Early methods involving thin-layer and gas-liquid chromatography, which in circumstances where the ratio of tocopherol to other lipids is very low may lead to losses, have now largely been superceded by high-performance liquid chromatography (HPLC), which has enabled a dramatic improvement in the ease, accuracy and sensitivity of tocopherol analysis. A range of different adsorbents has been used successfully with either a normal or reversed-phase mode of operation. Although α-tocopherol has the highest biological activity and may be of greatest interest to the investigator, other tocopherols form a significant part of the dietary intake and contribute to the total vitamin E activity of foods. Normal phase chromatography is usually employed for the separation of α-, β-, γ- and δ-tocopherols and a variety of modifying solvents, such as di-*iso*-propyl ether, diethyl ether, isopropanol, methanol and methyl-*t*-butyl ether, have been used to achieve separation in the main solvent which is usually hexane.

(c) **Detection and quantitation.** The tocopherols may be detected in the eluate from HPLC columns by taking advantage of their ultraviolet absorption in the region of 294 nm. However, this method is lacking in sensitivity and the method of choice, which greatly increases the sensitivity and specificity of detection, is to use a fluorescence method which also eliminates the spurious absorption of non-fluorescent compounds that would be detected by ultraviolet absorption in the same wavelength range as the tocopherols. The excitation wavelength for fluorescence detection is 220 nm, and the emission detection wavelength is 360 nm. Routine analysis of amounts of tocopherol around 1.0 ng is possible with a detection limit of about 0.25–0.5 ng depending on circumstances.

6.1.2. Recommended procedures

(a) **Serum.** The analysis of serum or plasma (and other body fluids) may be achieved by direct extraction of the tocopherols followed by HPLC analysis. Several such methods have been described (Bieri et al., 1979; McMurray and Blanchflower, 1979; Vatassery and Hagen, 1979). The following procedure is a similar method utilized by McCarthy and Diplock (unpublished data).

PROCEDURE. 1 ml of serum or plasma is extracted once with 4 ml hexane following the addition of 2 ml ethanol. The hexane layer is separated by centrifugation at $1500 \times g$ for 5 min and the hexane evaporated under N_2 at 70°C; the residue is dissolved in 50 μl spectroscopic grade hexane. The tocopherols are determined by HPLC by the method given in section (b) below.

(b) **Animal tissues and food products.** The amount of material to be analysed will depend on the expected tocopherol content; the method that follows is that of Buttriss and Diplock (1984) and it is valid for tissues containing normal amounts of tocopherol; the sample size should be increased for tissues from vitamin E-deficient animals.

PROCEDURE. The tissue sample (0.25–0.5 g) is homogenized at 0°C in 2 ml ethanol containing 1% pyrogallol, using a Polytron tissue homogenizer (Kinematica, GmBH), taking care to cool the container in crushed ice to prevent heating of the mixture during homogenization. An homogenization time of 0.5–1 min is usually sufficient, but tissues such as skeletal muscle may require longer. A saturated aqueous solution of KOH (0.3 ml) is added and the mixture heated in a waterbath at 70°C for 30 min with occasional shaking. The tocopherols in the mixture are extracted, following the addition of 1 ml of water, with 4 ml hexane with vigorous mixing on a vortex mixer. The mixture is centrifuged at 1500 × g for 5 min, the upper hexane layer removed and the residue re-extracted with 2 ml hexane. The hexane extracts are washed twice with 3 ml water using gentle agitation to avoid emulsion formation and the final hexane supernatant fraction, after the layers have been separated by centrifugation, is evaporated to dryness at 70°C under nitrogen. The residue is re-dissolved in 300 μl spectroscopic grade hexane.

HPLC

Column	:	Micropak Si5 (150 × 4.5 mm) or Lichrosorb Si60 (150 × 4.8 mm)
Solvents	:	HPLC grade hexane and HPLC grade methyl-t-butyl ether
Flowrate	:	2.0 ml/min
Conditions	:	350 atm at ambient temperature
Injection	:	30–100 μl in a 100 μl injection loop using a back-pressure restrictor
Detection	:	ultraviolet at 294 nm, *or*, for much greater

⟶

Fig. 6.1. HPLC separation of α-tocopherol (peak 1), γ-tocopherol (peak 2) and tocol (peak 3). **(a)** Using hexane and methyl-t-butyl ether (88:12), flow rate 2.0 ml/min, ambient temp., fluorescence detector, Lichrosorb Si60 column. **(b)** Using hexane and methyl-t-butyl ether (92:8), flowrate 2.0 ml/min, ambient temp., fluorescence detector, Lichrosorb Si60 column. **(c)** Using hexane and methyl-t-butyl ether (95:5), flowrate 2.0 ml/min, ambient temp., fluorescence detector, Lichrosorb Si60 column.

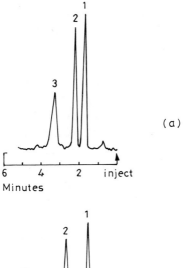

(a)

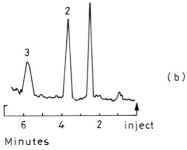

(b)

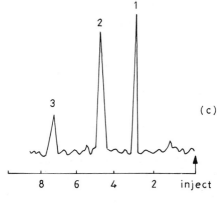

(c)

Solvent mixture : sensitivity and discrimination, fluorescence, 220 nm excitation, 360 nm emission may vary between 12 and 5% methyl-*t*-butyl ether in hexane. Fig. 6.1 gives typical separations of α- and γ-tocopherols and tocol: 1(a) was with 12% modifier, running time 4 min; 1(b) was with 8% modifier, running time 6 min; 1(c) was with 5% modifier, running time 8 min. If separation of tocopherols is required, 5% modifier is recommended; if rapid measurement of α-tocopherol only is required, 12% modifier may be used.

Recovery and validation : the recovery of a range of amounts of α-tocopherol (1–10 ng) added at the extraction stage should be greater than 98%. Further validation may be achieved by the use of γ-tocopherol; a linear relationship, with a high coefficient of variation, should be obtained when the amount of endogenous tocopherol recovered is plotted against increasing weights of tissue over a wide range of amounts of tissue.

This method is applicable to a wide range of tissues and should give recoveries 0.5–1.0 ng of α-, β- and γ-tocopherols better than 95% overall; greater sensitivity can be achieved by the use of an electrochemical detector. For measurement of tocopherols in foodstuffs it may be necessary to adapt the extraction procedure somewhat, and care must be taken to ensure the homogeneity of the sample.

(c) **Special procedure for high lipid-containing tissues.** For brain, which contains an unusually high concentration of lipid, it is necessary to adopt a somewhat different method of analysis. The method of Metcalfe et al. (1989) is a modification of that given above in section (b).

PROCEDURE. Approximately 250 mg of tissue is homogenized on ice in 10 vol of 75% ethanol in water using a Polytron tissue homogenizer (Kinematica, GmBH) for 1 min at the maximum speed. 12.5 vol of ice-cold hexane (HPLC grade) are added and the mixture vortex mixed for 30 s. Following centrifugation at $1500 \times g$ for 5 min, the supernatant hexane fraction is removed and may be stored for up to 48 h at $-20°C$ before analysis without loss of tocopherols. 101 μl of sample are used for HPLC. 10 μl of sample are used for HPLC.

HPLC
Column   : Jones Chromatography Silica (250 × 5 mm)
Solvents   : HPLC grade hexane containing 1% HPLC grade methanol; the methanol is dried by the addition of a small number of 4 Å molecular sieves
Flowrate   : 2.0 ml/min
Conditions : 350 atm at ambient temperature
Injection   : 10 μl in a 10 μl injection loop using a back-pressure restrictor
Detection   : fluorescence detection as given in method (b) above.

The authors state that the detection limit for α-tocopherol was of the order of 1.0 ng at a signal to noise ratio of 5, and the recovery of α-tocopherol added prior to homogenization was >90%; the coefficient of variation of the method was 4.7%.

6.2. Measurement of carotenoids

Current interest in the carotenoids as putative inhibitors of carcinogenesis (Connett et al., 1989) stems from the evidence that carotenoids can function as quenchers of singlet oxygen (Anderson and Krinsky, 1973) and oxygen-centred free radicals (Krinsky and Deneke, 1982).

The analysis of carotenoids in a wide range of biological materials can be readily and accurately carried out provided that care is taken to prevent oxidation, and isomerization, of the pigments by light. It is normal therefore to work in subdued light, preferably with amber glassware, to keep temperatures as low as is consistent with achieving proper extraction of the pigments, and to prevent excessive exposure to oxygen by storing and evaporating all solutions under N_2. The methodology for analysis of carotenoids falls into two stages:

(a) extraction of the pigment which may involve simple solvent extraction, or, in difficult cases, may require digestion with alkali. The procedure described below is based on that of Krinsky and Welankiwar (1984).

(b) measurement of the extracted carotenoids following separation by HPLC.

6.2.1. Procedure and HPLC

PROCEDURE: 1 g sample (tissue, foodstuff etc.) is homogenized in a tissue homogenizer or micro-blender with 6 ml of 70% (v/v) ethanol/water mixture. The carotenoid pigments are extracted with two successive 5 ml portions of petroleum ether with vortex mixing; emulsification is prevented by the addition of 2 ml of a solution (19%, w/v) of NaCl in water, and the layers are separated by centrifugation at $1500 \times g$ for 5 min. The resultant petroleum ether extracts are combined, evaporated to dryness by warming to 40°C in a stream of N_2, and the residue dissolved in HPLC-grade hexane. All procedures should be carried out in subdued light in amber glassware, and it is preferable that the subsequent HPLC analysis should be carried out immediately. In the case of difficult tissues such as skeletal muscle or hard materials that are not readily homogenized, it may be necessary to carry out a preliminary digestion procedure. The sample (1 g) is digested in an amber flask in 6 ml 6% KOH (w/v) at 40°C for between 10 min and 1 h. The extraction procedure follows that described above and the petroleum ether extracts are washed with successive 10 ml portions of water or dilute NaCl solution until the washings are alkali free.

HPLC

Column	:	Lichrosorb RP-18, 10 μl (250 × 4.6 mm)
Solvents	:	all solvents of HPLC grade and degassed.

System 1: acetonitrile/methanol (85:15).
System 2: hexane/methanol (25:75). System 1 is used at the beginning of the run and is changed to system 2 after 11 min.

Flowrate	:	2.0 ml/min
Conditions	:	350 atm at ambient temperature
Injection	:	10–20 μl in a 20 μl injection loop using a back-pressure restrictor.
Detection	:	visible at 450 nm.

A typical separation of spinach carotenoids supplemented with zeaxanthin is given in Fig. 6.2; this is derived from Krinsky and Welankiwar (1984). The identity of the peaks may be established by collecting

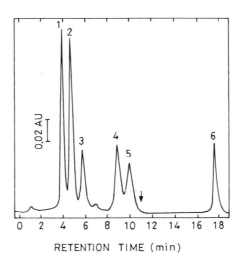

RETENTION TIME (min)

Fig. 6.2. HPLC separation of spinach carotenoids with added zeaxanthin. A 101 m Lichrosorb RP-18 column was eluted for 11 min at 2.0 ml/min with acetonitrile/methanol (85:15); the solvent was changed to hexane/methanol (25:75) for the remainder of the run. Detection 450 nm. Peak 1: neoxanthin; Peak 2: violaxanthin; Peak 3: antherxanthin; Peak 4; lutein; Peak 5: zeaxanthin; and Peak 6: β-carotene.

the fractions from HPLC and subjecting them to spectrophotometric identification and measurement. Identification depends on the spectral properties of the individual carotenoid pigments; the individual absorption spectrum depends on the number and arrangement of conjugated double bonds as well as on other functional groups. Quantitation may be achieved by reference to published $E_1\,_{cm}^{1\%}$ values (Davies, 1976; DeRitter and Purcell, 1981).

6.3. Measurement of antioxidant enzymes

6.3.1. Glutathione peroxidase

Glutathione:hydrogen-peroxidase oxidoreductase (EC 1.11.1.9) is a selenoprotein enzyme that catalyses the reduction of a large number of hydroperoxides ranging from H_2O_2 to a spectrum of organic hydroperoxides (ROOH).

$$H_2O_2 + 2GSH \rightarrow GSSG + 2H_2O$$
$$ROOH + 2GSH \rightarrow GSSG + ROH + H_2O$$

There is some overlap in activity with other enzymes, in particular with catalase, which can only catalyse the reduction of H_2O_2, and the glutathione S-transferases which can only catalyse the reduction of lipid hydroperoxides. Measurement of glutathione peroxidase is fraught with difficulty, particularly when crude tissue samples are to be measured, because other enzyme activities that mimic those of glutathione peroxidase can give falsely high values because other substances such as haemoglobin may have pseudoperoxidase activity and because the enzyme has very unusual kinetics. With purified enzyme, ping-pong kinetics have been established with infinite limiting maximum velocities and variable Michaelis constants. It follows that the usual requirement for saturating concentrations of the substrates involved cannot be achieved in the case of this enzyme. Compromises are therefore necessary in the concentration of substrate to be employed in the assay and in the definition of the unit of enzyme activity. Although methods have been devised for the measurement of GSH utilization by the enzyme, and of H_2O_2 consumption, these are difficult and have now largely been replaced by an assay based on that of Paglia and Valentine (1967) in which the product GSSG is used to drive the oxidation of $NADPH + H^+$ catalysed by glutathione reductase:

$$GSSG + NADPH + H^+ \rightarrow 2GSH + NADP^+$$

The oxidation of $NADPH + H^+$ is measured spectrophotometrically and, under appropriate conditions, may be used to measure the activity of glutathione peroxidase. The procedure which follows is that described by Flohé and Gunzler (1984).

6.3.1.1. Procedure

REAGENTS: 0.1 M potassium phosphate buffer, pH 7.0, containing 0.1 mM EDTA.

2.4 U/ml glutathione reductase (from yeast prepared daily by dilution from a reliable commercial source).

10 mM GSH in water

1.5 mM NADPH in 0.1% $NAHCO_3$

12 mM t-butylhydroperoxide

1.5 mM H_2O_2 in water

Enzyme sample, to contain 0.05–1.0 U_k/ml

0.1 M phosphate buffer pH 7.0.

ASSAY: The following are pipetted into a semi-micro cuvette thermostatted at 37°C: 500 μl 0.1 M phosphate buffer pH 7.0, 100 μl enzyme sample, 100 μl glutathione reductase (0.24 U); exactly 100 μl 10 mM glutathione solution is added and the hydroperoxide independent oxidation of NADPH measured for 3 min to give a baseline at 340 nm. The reaction is started by the addition of 100 μl of the solution of either t-butyl hydroperoxide or H_2O_2, prewarmed to 37°C. The decrease in absorbance is further monitored for 5 min. The non-enzymic reaction rate is assessed by replacing the glutathione peroxidase-containing enzyme sample by buffer.

Calculation of enzyme activity The decrease in NADPH concentration, Δ(NADPH)/min is calculated from the linear slope of the graph obtained and the appropriate absorption coefficient for NADPH. The true glutathione peroxidase-dependent reaction rate is obtained by subtracting the value of the non-enzymic and hydroperoxide-independent apparent rates. Glutathione peroxidase activity is frequently given simply in terms of Δ(NADPH)/min under standard conditions. A calculation of an activity value, A, that is likely to be more universally applicable, is given in by Flohe and Gumzler (1984).

Special conditions For tissue such as liver which may contain substantial amounts of glutathione *S*-transferases, H_2O_2 only should be used as substrate. An approximate measurement of glutathione *S*-transferase activity may be obtained by measuring the apparent glutathione peroxidase activity with *t*-butyl hydroperoxide and H_2O_2 and subtracting one from the other. It is also necessary, when H_2O_2 is used as substrate to block the activity of catalase by adding 1 mM sodium azide to the assay mixture. Traces of methaemoglobin will falsify the measurement of glutathione peroxidase. Thus, if the enzyme is to be measured in red blood cells (or in plasma that contains even traces of haemolysed red blood cells) the conversion of haemoglobin to cyanomethaemoglobin prior to the assay should be carried out. This is done by treating the sample with a 1–2 -fold excess of hexacyanoferrate(III) and a 12-fold excess of cyanide over haem concentration. In these circumstances *t*-butyl hydroperoxide should be used as substrate.

6.3.2. *Glutathione S-transferases*

The glutathione *S*-transferases (GSTs) are a group of enzymes that catalyses a range of reactions that are important in physiological detoxification mechanisms. Three main reaction types have been identified.

Type 1: The enzyme catalyses the nucleophilic attack of GSH on an electrophilic centre in the substrate to produce a stable GSH conjugate:

$$R\text{-}X + GSH \rightarrow R\text{-}SG + XH$$

Typical reactions in this category occur with 1-chloro-2,4-dinitrobenzene, 1,2-chloro-4-nitrobenzene, ethacrynic acid, *trans*-4-phenyl-3-buten-2-one and a range of hydroxylated derivatives of polycyclic hydrocarbons.

Type 2: The enzyme catalyses a similar reaction to Type 2 initially, but the product is subject to further non-enzymic attack by GSH to produce a reduced substrate and oxidized GSH:

$$R\text{-}X + GSH \rightarrow R\text{-}SG + XH \text{ (enzymic)}$$
$$R\text{-}SG + GSH \rightarrow RH + GSSG \text{ (non-enzymic)}$$

This category of reaction is typified by the catalytic reduction by GSTs of organic hydroperoxides thus:

$$ROOH + GSH \rightarrow RO\text{-}SG + H_2O$$
$$RO\text{-}SG + GSH \rightarrow ROH + GSSG$$

Type 3: This is an entirely different function in which certain forms of the GSTs catalyse the isomerization of steroids, thus:

$$\Delta^5\text{-androstene-3,17-dione} + GSH$$
$$\downarrow$$
$$\Delta^4\text{-androstene-3,17-dione} + GSH$$

GSH in this instance functions only as a coenzyme.

The activity of the GST-peroxidase (Type 2) is difficult to measure directly and individually, since the observed activity in tissue fractions will be the sum of the glutathione peroxidase and GST activities. In practice, therefore, the GST-peroxidase activity may be obtained by measuring the glutathione peroxidase activity first with H_2O_2 as substrate (which gives a measure only of the glutathione peroxidase activity) and secondly with a hydroperoxide, such as t-butyl or cumene hydroperoxide, as substrate (which gives the sum of the glutathione peroxidase and GST-peroxidase activities). The GST-peroxidase activity may then be obtained by subtracting the first value from the second.

It is often desirable, in order to obtain a picture of the entire GST activity of a tissue or tissue fraction, to measure the enzyme with several different substrates; this can conveniently be done by a modification of the spectrophotometric method of Habig and Jakoby (1981).

6.3.2.1. Procedure

Assay conditions for a range of different substrates are set out in Table 6.1; substrate and GSH concentrations given are final concentrations in the complete reaction mixture.

The formation of the GSH conjugate (or the Δ^4 isomer of androstene-3,17-dione) is monitored at 30°C in a jacketed spectrophotometer micro-cell. The GSH-containing sodium phosphate buffer (0.1 M, 0.9 ml) of appropriate GSH concentration, is added first to the cuvette followed by the substrate in 0.05 ml ethanol (or methanol for Δ^5-androstene-3,17-dione). The reaction is initiated by the addition of the enzyme preparation in the appropriate sodium phosphate buffer and

TABLE 6.1
Assay conditions for a range of different substrates

Substrate	Substrate conc. (mM)	GSH conc. (mM)	pH	λ_{max} (nm)	$\Delta\varepsilon$ (mM$^{-1}\cdot$cm^{-1})
1-Chloro-2,4-dinitrobenzene	1.0	1.0	6.5	340	9.6
1,2-Chloro-4-nitrobenzene	1.0	5.0	7.5	345	8.5
Ethacrynic acid	0.2	0.25	6.5	270	5.0
trans-4-Phenyl-3-buten-2-one	0.05	0.25	6.5	290	24.8
Δ^5-Androstene-3,17-dione	0.07	0.1	8.5	248	16.3

the progress of the reaction is monitored at the wavelength indicated in the table; a blank reaction is done using buffer instead of the enzyme preparation. The GST activity of the preparation is calculated using the appropriate absorption coefficient given in the table and the activity expressed as μmol conjugate (or $4\varDelta$ isomer) formed per min.

6.3.2.2. Further separation and purification

The resolution of the GSTs into their constituent isoforms, and the separation of each isoform into the constituent subunits, is a complex procedure beyond the scope of the present work. Initial purification of GSTs from tissue fractions is best achieved by the affinity chromatographic method of Guthenberg and Mannervik (1979). The affinity-purified factions may be further purified using fast protein liquid chromatography (FPLC) electrofocussing (Alin, 1984), although some modification of the gradient system and flow rates used by these authors is required to resolve all the GSTs, particularly the acidic isoforms. Resolution of the subunits of each dimer obtained from the above separation is best achieved by an adaptation of the sodium dodecyl sulphate-polyacrylamide gel electrophoresis (SDS-PAGE) technique of Laemmli (1970).

6.3.3. Catalase

Catalase has a dual functional role; a true catalytic role in the decomposition of hydrogen peroxide:

$$2H_2O_2 \rightarrow 2H_2O + O_2$$

and a peroxidic role in which the peroxide is utilized to oxidize a range of H donors (AH_2) such as methanol, ethanol and formate:

$$AH_2 + H_2O_2 \rightarrow A + 2H_2O$$

In each case an active enzyme-H_2O_2 complex is formed initially followed by an exceedingly rapid second stage in which a second mole-

200   TECHNIQUES IN FREE RADICAL RESEARCH

cule of H_2O_2 serves as H donor for the enzyme-H_2O_2 complex. The kinetics of the enzyme do not obey normal rules since there is strong inactivation of the enzyme above an H_2O_2 concentration of 0.1 M; determination of K_s is therefore impossible. The enzymic decomposition of H_2O_2 is first-order, the rate being proportional to the concentration of peroxide. The assay is carried out at a quite low concentration of H_2O_2 (≈ 0.01 M) so as to avoid a rapid fall in the initial rate of the reaction; during about 1.0 min at H_2O_2 concentrations in the range 0.01–0.05 M, the decomposition of H_2O_2 is first order. The following method is that of Aebi (1984).

6.3.3.1. Procedure
REAGENTS: Phosphate buffer, 50 mM, pH 7.0 (to prepare dissolve 6.81 g KH_2PO_4 and 8.90 g $Na_2HPO_4\cdot 2H_2O$ in distilled water and dilute each to 1 litre; mix in the proportions 1:1.5 (v/v), respectively). Hydrogen peroxide 30 mM: dilute 0.34 ml of 30% hydrogen peroxide with phosphate buffer to 100 ml.

FOR BLOOD: heparinized venous blood is centrifuged and the upper layer is removed. Wash the erythrocyte sediment three times with 0.9% (w/v) NaCl solution. Haemolyse the washed red cells by adding 4 parts (v/v) of distilled water per volume of packed cells to give a stock haemolysate solution (approx. 5% (w/v)). For assay, dilute the stock solution 1:500 with the phosphate buffer immediately before the assay is to be carried out and determine the haemoglobin content of the solution by the method of Drabkin. The catalase activity is expressed per unit of haemoglobin.

FOR TISSUES: the spectrophotometric method can be used for tissues such as liver and kidney which have a high content of catalase provided all cells and debris can be removed to give a solution that is clear and only slightly coloured. A stock homogenate should be made in buffer containing 1% Triton X-100 (1 g tissue + 9–19 ml buffer), and further dilutions made to reach an appropriate concentration of enzyme which can only be determined by trial. If such dilution is not possible due to low catalase activity in the sample it may not be possible to obtain satisfactory ultra-violet measurement due to the absorption of residual Triton X-100. In such cases it may be possible to achieve the desired end by replacing the Triton X-100 by 0.01% (w/v) digitonin or 0.25% (w/v) sodium cholate. The catalase activity of samples is usually referred to a wet weight basis.

ASSAY: place 2.0 ml of haemolysate or enzyme preparation in a 3.0 ml cuvette and set the spectrophotometer to 240 nm at room temperature. Add 1.0 ml of the H_2O_2 solution, mix well and follow the decrease in absorbance for about 30 s: the initial absorbance should preferably be around 0.5.

CALCULATION OF ACTIVITY: many different methods of defining and calculating units of enzyme activity have been used and these are reviewed by Acbi (1984). The use of the first order rate constant (k) has merit as a means of expressing enzyme activity and this may be calculated as follows:

$$k_{15} = 0.153(\log A_1/A_2)$$

where $A_1 = A_{240}$ at $t = 0$, k_{15} is k at the 15-s time interval, and $A_2 = A_{240}$ at $t = 15$ min. Appropriate further calculations may then be applied to calculate k_{15}/g Hb or k_{15}/g tissue.

6.3.4. Superoxide dismutase

Superoxide dismutase (SOD; EC 1.15.1.1) is the term used for a number of metalloproteins that catalyse the following reaction:

$$2O_2^- + 2H^+ \rightarrow H_2O_2 + O_2$$

Assay of SOD is difficult because of the free radical nature of its substrate ($O_2^{\cdot-}$) which must of necessity be generated within the assay system and which cannot be measured directly by simple analytical means. Indirect measurement, using an indicator of $O_2^{\cdot-}$ concentration, is therefore used as a means of measuring the activity of SOD in the system. This method is fraught with sources of error, since any other entity which scavenges $O_2^{\cdot-}$ in the system, or changes the rate of $O_2^{\cdot-}$ formation, or reacts with the indicatory system, will vitiate the result obtained. Thus, the use of the assay to be described here is reliable for a purified enzyme system and becomes progressively less reliable the less pure the enzyme preparation is. These potential errors, and some means of overcoming them, are discussed in detail by Flohé and Ötting (1984). The method given is based on that originally described by McCord and Fridovich (1969); the xanthine-xanthine oxidase system is used to generate $O_2^{\cdot-}$ and the reduction rate of cytochrome by $O_2^{\cdot-}$ is monitored at 550 nm. The inhibition of this reduction when the SOD containing preparation is added is used as a measure of the activity of SOD.

$$\text{Xanthine} + O_2 \xrightarrow{\text{XO}} O_2^{\cdot-}$$
$$O_2^{\cdot-} + \text{cytochrome } c(Fe^{3+}) \longrightarrow O_2 + \text{cytochrome } c\ (Fe^{2+})$$

and

$$2O_2^{\cdot-} + 2H^+ \xrightarrow{\text{SOD}} H_2O_2 + O_2$$

6.3.4.1. Procedure
REAGENTS: 50 mM potassium phosphate buffer, pH 7.8, containing 0.1 mM ethylene diamine tetraacetic acid, 0.76 mg xanthine in 10 ml 0.001 M sodium hydroxide solution, 24.8 mg cytochrome c in 100 ml phosphate buffer, pH 7.8.
Solution A is prepared by mixing 10 ml of the xanthine solution with 100 ml of the cytochrome c solution: the mixture is stable at 4°C for about 3 days.
Solution B is a solution of xanthine oxidase (Grade I, Sigma) in the EDTA-containing buffer, pH 7.8, to give an activity of about 0.2 U/

ml, which will give a rate of reduction of cytochrome c of 0.025 absorbance units/min in the absence of SOD.

ASSAY: since one unit of SOD activity is defined as the amount of enzyme that inhibits by 50% the rate of reduction of cytochrome c under specified conditions, it is necessary to try several different dilutions of the enzyme preparation. Solution B should be kept at 4°C and solution A warmed to 25°C; a thermostatted spectrophotometer cell at 25°C should be used at 550 nm; for maximum sensitivity the spectrophotometer should be set to the observed maximum absorbance when a portion of solution A is reduced with a few crystals of dithionite. 2.9 ml of solution A is then placed in a 3 ml cuvette and 50 μl of the enzyme sample is added with mixing. The reaction is started by adding 50 μl of solution B with further mixing and the change in absorbance at 550 nm is monitored. The enzyme sample should be replaced by water or by several standard SOD solutions to obtain a blank value, which should be subtracted, and a range of standard curves. Plots of $1/\Delta E$ min^{-1} for the standard enzyme are used to determine the activity of the unknown enzyme preparation; the ΔE min^{-1} value is obtained from the linear part of the curve.

SPECIAL CONDITIONS: the above method is valid for a pure enzyme preparation, but cannot give entirely reliable measurements for impure samples. Interfering reactants in the medium may be allowed for by carrying out recovery experiments with a range of amounts of pure SOD added to the test enzyme preparation. Dialysis of the enzyme preparation will eliminate small molecules that may interfere, like ascorbate, reduced glutathione and catecholamines. The addition of 2 μM cyanide may be used to block peroxidases, which has only a minimal effect on the activity of Cu/Zn-SOD. Alternatively 10^{-5} M azide may be used to block peroxidases without effect on Cu/Zn-SOD.

Assay of Cu/Zn-SOD and Mn-SOD or Fe-SOD separately may be achieved as follows: total SOD activity is measured as described above and the assay is then repeated following the addition of 2 mM cyanide which completely blocks the activity of the Cu/Zn-SOD. The activity of the Cu/Zn enzyme and the Mn enzyme (in mammalian tissue preparations) may then be obtained by difference.

6.4. Measurement of selenium in biological samples

The method described is an adaptation by Coker (1978) of the original method of Olson (1975). Selenium is extracted by digestion of the biological sample in a mixture of concentrated HNO_3, and perchloric acid. The resultant mixture which contains the selenium as SeO_3^{2-} is reacted with diaminonaphthalene (DAN) and the diazoselenol is extracted into cyclohexane.

2,3-diaminonaphthalene   naphthylo-1,5-diazoselenol

The diazoselenol gives a bright red fluorescence when irradiated at 385 nm and the emission may be measured at 515 nm as the basis of the assay of selenium.

6.4.1. Procedure

REAGENTS:

1. DAN reagent. 2,3 diaminonaphthalene is dissolved in 0.1 M HCl. The reagent is ground to a fine powder before weighing, dissolved and the solution placed in a 250 ml separating funnel. 100 ml cyclohexane is added and the mixture shaken gently and allowed to separate in the dark. The upper cyclohexane layer is discarded and the aqueous layer re-extracted with a further 150 ml portion of cyclohexane. The prepared reagent is stored in an amber bottle covered with aluminium foil with a small amount of cyclohexane on the surface of the liquid to exclude air.
2. Hydroxylammonium/EDTA reagent. Dissolve 9 g of EDTA and 25 g hydroxylammonium chloride in 1 litre water.
3. Cresol red. Dissolve 20 mg cresol red in 100 ml 0.001 M NaOH solution.

4. Stock selenium solution. Dissolve 10 mg elemental selenium (00.00% purity) in 0.8 ml concentrated nitric acid with warming and dilute to 150 ml with water.

5. Working selenium solution. Dilute 400 μl of stock solution to 200 ml with 0.1 M hydrochloridic acid.

All glassware should be rinsed with 10% nitric acid after washing, to remove traces of detergent and the glassware rinsed with deionized water before use.

ASSAY

Triplicate samples and a range of aliquots of standard selenium solution are placed in 15 ml stoppered tubes and 1.25 ml concentrated nitric acid added, and the mixture left to stand overnight. By the following day biological samples should be well digested; 0.5 ml perchloric acid and a few anti-bumping granules are then added. The samples and standards are then carefully heated until all the HNO_3 has been removed at which time white fumes of perchloric acid appear; this procedure takes about 30 min. A number of samples may be heated simultaneously on a mini-Kjeldahl heater, the tops of the tubes being inserted in a glass manifold within which a negative pressure is maintained with a water pump to disperse the HNO_3 fumes. *Great care must be taken to ensure that the mixtures are not overheated,* which may volatilize the selenium, and it is vital to stop heating at once when the white fumes of perchloric acid appear. The selenium is oxidized under these conditions to SeO_3^{2-} and, when the procedure is completed, 2.0 ml of 10% (v/v) hydrochloric acid is added while the samples are still hot. After cooling 2.5 ml of the hydroxylammonium-EDTA reagent is added followed by 4 drops of the cresol red indicator. The samples are then titrated to pH 2.0 with 10% HCl the end point being a peachy-red colour. Deionized water (2.5 ml) and 2.5 ml of the DAN reagent are now added and the samples covered loosely (glass marbles are ideal) and placed in a water bath at 50°C for 30 min. They are then cooled in the dark and 1 ml cyclohexane is added, the tubes shaken vigourously for 20 s to extract the selenium-DAN complex, and the upper cyclohexane layer is allowed to separate for 15 min in the dark.

The spectrofluorimeter is set to an excitation wavelength of 365 nm and an emission (detection) wavelength of 515 nm. Using a cyclohexane blank, the emission reading is set to 100% with the standard sample containing the largest amount of selenium. All samples and standards are read at the same settings and a standard curve is constructed from which the selenium content of the unknown samples is determined. It is particularly important to use a range of standards each time that an assay is carried out since it has been found that there is some day to day variability in the intensity of fluorescence of the selenium-DAN derivative.

Special procedures and alternatives
The method described may be adapted to a wide range of biological samples and food-stuffs. It may be found necessary to warm the sample in concentrated HNO_3 to effect complete digestion, and also to use rather more HNO_3 than that recommended above.

Hydride generation atomic absorption spectrometry and electrothermal atomisation atomic absorption spectrometry may also be employed for the measurement of selenium in biological samples. A comparison was made (MacPherson et al., 1988) of these methods with the fluorimetric method described here and all three methods were found to give accurate, reproducible results when samples of plasma and urine with certified selenium contents were analysed.

CHAPTER 7

Detection of protein structural modifications induced by free radicals

7.1. Introduction

All the constituent amino acid side-chains in proteins are susceptible to free radical attack, but some are more vulnerable than others, as discussed in Chapter 2. Thus, exposure of proteins to free radical-generating systems may induce tertiary structural changes as a consequence of modifications to individual amino acid side-chains. As secondary structure is stabilized by hydrogen bonding between peptide groups, interactions of radical species with the polypeptide backbone and interference with the functional groups of the peptide bonds may cause secondary structural modifications.

In this chapter, methods for measuring and detecting free radical-induced modifications to protein structure will be described. Techniques for investigating primary structural changes include amino acid analysis, as well as methods for measuring specific amino acid side-chains including the fluorescamine assay for primary amino groups, intrinsic fluorescence detection of aromatic side-chains and the measurement of reduced thiol side-chains. Infrared spectroscopic techniques are also described and applied to assess the secondary structure, as well as structural modifications to specific amino acid side-chains. Disruption of the secondary structure may also occur under certain conditions of free radical attack at the α-carbon atom of the peptide bond leading to an increase in the carbonyl content of the protein. This method of assessing increased carbonyl content is also described.

7.2. The use of infrared spectroscopy to determine structural changes

Infrared spectroscopy is a well-established technique for the study of protein structural modifications. Proteins show several characteristic absorptions termed the 'amide bands' corresponding to particular types of structure within proteins (Koenig and Tabb, 1980; Susie and Byler, 1983). Spectra are usually presented as absorbance versus wavelength or, more commonly, wavenumber (cm^{-1}), the reciprocal wavelength. Table 7.1 shows the characteristic IR absorption bands of the peptide linkage. Of these bands, the most useful are the amide bands I and II, lying in the region 1900 to 1400 cm^{-1}, which are sensitive to changes in secondary structures. In addition several amino acid side-groups, including tyrosine, aspartate and glutamate show characteristic absorption bands. The intense absorption of liquid water obscures much of the IR spectrum of biomolecules in aqueous systems. However, accurate computer subtraction of the water bands greatly reduces this problem, though care needs to be taken because

TABLE 7.1
Characteristic infrared absorption bonds of the peptide linkage

Conformation	Amide I (cm^{-1})	Amide II (cm^{-1})
α-Helix	1652 (s)	1546 (s)
	1646 (w)	1516 (w)
Random coil	1656	1520
Anti-parallel chain β-sheet	1632 (s)	1530 (s)
	1690 (w)	1510 (w)
Parallel chain β-sheet	1632 (s)	1530 (s)
	1648 (w)	1550 (w)

s = strong; w = weak.

Suspension of soluble proteins in ^2H_2O shifts the amide II band (principally N-H bending) to frequencies near 1450 cm^{-1} (amide II').

the water bands themselves are solute sensitive (Chapman et al., 1980). This greatly facilitates the assessment of structural modifications (for reviews, see Lee and Chapman, 1986; Haris et al., 1989).

7.2.1. The principle

The amide I and II bands of peptide bonds in proteins are particularly sensitive to changes in secondary structure and are thus the most commonly used vibrational modes in conformational analysis. As shown in Table 7.1, the amide I region comprises vibrations due mainly to $C=O$ stretching modes with some contribution from N-H bending and C-N stretching, whereas the amide II frequencies correspond mainly to N-H bending with a large contribution from C-N stretching modes.

These vibrations are influenced by the secondary structure of the protein, since this involves protein folding with hydrogen bonding between peptide bonds. Characteristic amide I and II frequencies for various polypeptide configurations in H_2O and ^2H_2O solvents are presented in Table 7.2 showing that $^1H-^2H$ exchange results in a shifting of amide I frequencies to lower values. Because the amide II band involves the N-H unit, replacing ^1H by ^2H causes a drastic shift. Any residual absorption in the normal region must be due to other vibrational modes or to residual $N-^1H$ units that have not exchanged with ^2H_2O.

The combination of infrared spectroscopy and hydrogen–deuterium exchange is a powerful technique for revealing small differences in protein secondary structure. Few proteins are composed solely of one type of structure, therefore several amide I and amide II frequencies may contribute to any amide I and II band. It is often difficult to resolve all of these frequencies in the difference spectrum, since some of the peaks have bandwidths which are smaller than the amide I or amide II bandwidth and are thus effectively hidden within the main peak. To resolve overlapping bands, second derivative spectra may be generated using a computer programme. The resultant spectrum is presented as absorbance/(wavenumber)^2 versus wavenumber.

TABLE 7.2

Amide I vibrations – characteristic amide I frequencies in H_2O and ^2H_2O solvents

Conformation	Amide I (cm^{-1})	
	H_2O	^2H_2O
α-Helix	1652 (s) 1646 (w)	1650 (s) 1644 (w)
Random coil	1656	1643
Anti-parallel chain β-sheet	1632 (s) 1690 (w)	1632 (s) 1675 (w)
Parallel chain β-sheet	1632 (s) 1648 (w)	1630 (s) 1645 (w)

s = strong; w = weak.

Such derivative techniques can reveal small changes in spectra and hence in protein secondary structure which will be picked up in the second derivative spectrum of the protein as gain or loss of characteristic band intensities (Susi and Byler, 1983; Alvarez et al., 1987).

7.2.2. The spectrometer

Although standard IR spectrometers are used for studying the amide bands, FTIR spectrometers are more accurate and reliable. FT-IR spectrophotometers are based upon the Michelson interferometer. A typical instrument (Fig. 7.1) comprises an optical bench housing the interferometer, sample, infrared source and detector, coupled to a computer, which controls the spectral scanning, analysis and data processing (for review see Griffiths, 1980).

The Michelson interferometer consists simply of two mutually perpendicular plane mirrors one of which is fixed and the other able to move at 90° to its plane. A semi-reflecting film or 'beamsplitter' bisects the plane of the two mirrors. An ideal beamsplitter has zero absorption of incident light and 50% reflectance. Collimated, incident IR radiation is split by the beamsplitter, so that 50% of it is reflected to one of the mirrors and 50% of it is transmitted to the other. On reflection from the mirrors, the beams recombine at the beamsplitter. Again 50% of the reflected beam from each mirror is reflected back to the mirrors and 50% of each beam passes through the beamsplitter and is termed the transmitted beam. It is this beam which passes through the sample and is picked up by the detector. The moving mirror generates an interferogram which is also detected. The detector output is converted to digital form and stored in the computer. By means of a fast Fourier transform, the interferogram is converted to the normal absorbance versus wavenumber spectrum.

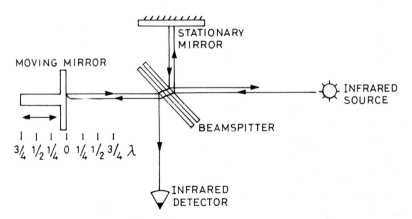

Fig. 7.1. Layout of the infrared spectrometer showing the Michelson Interferometer Optical System. An FTIR spectrometer's optical system requires two mirrors, an infrared light source, an infrared detector and a beamsplitter. The beamsplitter reflects about 50% of an incident light beam and transmits the remaining 50%. One part of this split light beam travels to a moving interferometer mirror, while the other part travels to the interferometer's stationary mirror. Both beams are reflected back to the beamsplitter where they recombine. Half of the recombined light is transmitted to the detector and half is reflected to the infrared source.

7.2.3. Procedure

7.2.3.1. Preparation of buffers

For the studies described here phosphate-buffered saline, pH 7.4, is used as solvent. For the deuteration studies, 0.312 g of NaH_2PO_4 and 0.1753 g NaCl are quickly dissolved in 10 ml ^2H_2O. To obtain a p^2H equivalent to a pH of 7.4, aliquots of NaO^2H were added until the meter reading was 7.0. (The electrodes of a pH meter are less sensitive to $^2H^+$ than to $^1H^+$, therefore the meter gives an underestimate of $^2H^+$ concentration. Overshoot of p^2H above 7.0 may be corrected using ^2HCl.) The solution is then made up to 20 ml with ^2H_2O and the p^2H checked. Since ^2H_2O absorbs water readily from the atmosphere, it is essential to work quickly and to minimize contact of the solution with the atmosphere.

7.2.3.2. Preparation of the sample

Sample concentrations > 20 mg/ml of protein are used, since lower concentrations do not give spectra with a good signal-to-noise ratio in aqueous solvent. Once prepared the solution is sealed in its container with parafilm and stored at 4°C.

7.2.3.3. The microcell and the loading of the sample

30 to 50 μl of sample is placed in a sealed microcell (Beckman FH-01 CFT) equipped with a 6 μm tin spacer. The choice of the window material is dictated by the requirements for transparency in the mid-infrared and minimal solubility in water. Calcium fluoride is the material of choice which is almost insoluble in water and is transparent in the region 5000–1000 cm^{-1}. For studies down to 450 cm^{-1} silver chloride may be used (with the disadvantages of both high reflection losses and expense). The pathlength of the cell for solution work must be selected to take account of the absorptivity of the solvent.

Loading of the sample is achieved either by injection through one of the points provided for this purpose or by placing the sample on the lower window which supports the spacer and carefully lowering the upper window to form a thin film of the sample without formation of air bubbles. The screws holding the window in place are tightened. Injection is usually used in those cases where the sample is available in quantities such that the dead volume of the cell is not a consideration. For samples of which only very small volumes are available, such as concentrated solutions of membrane proteins, the cell is filled by film formation. In this case it is essential that care is taken to eliminate air bubbles from the window area. The temperature of the sample in the spectrophotometer is maintained by means of a water jacket surrounding the cell in its position inside the machine. The spectrometer is continuously purged with dry air.

Cells are cleaned by rinsing with distilled water after complete disassembly. The components of the cell must be thoroughly dried before reloading a fresh sample.

7.2.4. The spectra

The sample is allowed to equilibrate for 15 min inside the instrument before scanning is commenced. The spectrometer is continually purged with dry air to avoid interference from water vapour absorption. 400 scans are averaged at a resolution of 4 cm^{-1}. The scanning is complete for each sample in 1 h.

Difference spectra are obtained by digital subtraction of a spectrum of the buffer spectrum from the corresponding sample spectrum using an interactive difference function (IDIFF) recorded on the same day under the same conditions as the sample spectrum. Subtraction is adjusted to give a flat baseline in the region extending from 1900 cm^{-1} to the base of the amide I peak. Difference spectra in the region of significance 1900 cm^{-1} to 1400 cm^{-1} are saved.

Examples of the use of IR spectroscopy to detect small changes in secondary structure or changes in primary structure are given for radical-modified caeruloplasmin and immunoglobulin G. As will be demonstrated, in some instances small significant changes in the secondary structure can be detected, in other instances the data do not indicate secondary structural modifications. Thus, a key break in the structure of the protein in the appropriate place might lead to dramatic changes in the secondary structure.

7.2.4.1. *Oxidatively modified caeruloplasmin*

A difference infrared spectrum of a control sample of caeruloplasmin after the subtraction of ^2H_2O is shown in Fig. 7.2, where the amide I maximum is centred at 1641 cm^{-1} (Haris et al., 1989). This band frequency is consistent with the presence of predominantly β-sheet structure in this protein with a proportion of α-helix.

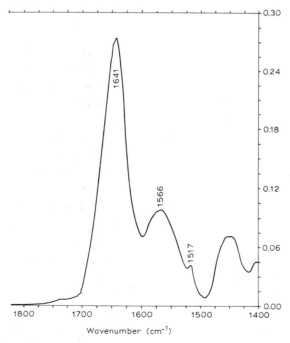

Fig. 7.2. Difference spectrum of caeruloplasmin after subtraction of ^2H_2O (Haris et al., 1989).

Fig. 7.3 shows the *second-derivative spectra* of control caeruloplasmin and of caeruloplasmin (20 mg/ml) treated with hydroxyl radicals generated from 0.5 mM Cu(II) and 5 mM hydrogen peroxide. These spectra were obtained for samples in ^2H_2O. Control caeruloplasmin has bands near 1637 cm^{-1} indicating β-structure with a relatively weaker band near 1652 cm^{-1} due to a small amount of α-helix/random coil structure. On exposure of caeruloplasmin to the hydroxyl radical-generating system, this band shifts to 1656 cm^{-1}, indicative of a transition to the disordered conformation, with no major shift in the original band indicative of the β-structure. Thus FTIR-spectroscopy allows sensitive demonstration of the loss of the small amount of α-

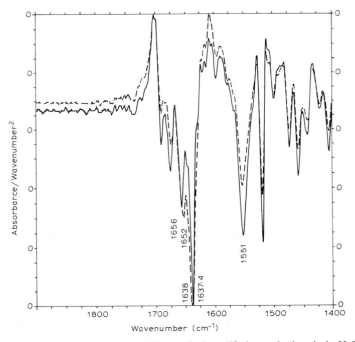

Fig. 7.3. Second derivative spectra of free radical-modified caeruloplasmin in H_2O saline. Control caeruloplasmin (- - - - -). Caeruloplasmin exposed to hydroxyl radicals (----) (Haris et al., 1989).

helical structure in this protein whilst not discernibly affecting the β-structure under these conditions of exposure to oxygen radicals.

7.2.4.2. Oxidatively modified immunoglobulin G

Fig. 7.4 demonstrates the difference spectrum of IgG in ^2H_2O/PBS buffer, p^2H 7.4, before and after exposure to hydroxyl radicals, after subtraction of buffer spectrum (Haris et al., 1989). The amide I maximum centred at 1639 cm^{-1} is consistent with a predominance of β-sheet structure. The absence of any changes in the amide I band for this example indicates that alteration in the secondary structure is unlikely. From these data, the major difference between the control and the samples exposed to oxygen radicals is the greater absorbance for the latter near 1587 cm^{-1}. Absorbance in this region is associated with ionized carboxylate groups (-$CO_2{}^-$), which increases on exposure of protein residues to oxygen radicals, hydrolysing to glutamate. It is conceivable that under these oxidizing conditions proline residues are converted to carboxylate side-chains.

Thus, it is important to realize that just because there is no change in the spectrum does not necessarily mean no change in the overall secondary structure of the protein. For example, a polypeptide chain scission may take place inducing a break in a disordered region, leaving individual separate portions of secondary structure intact.

7.2.5. Advantages and disadvantages of the technique

1. An advantage of this technique is that an external probe molecule is not required and the absorption of the protein groupings reflect their genuine environments. Techniques such as fluorescence are limited, to a certain extent, by the perturbation which added probes may induce.

2. The timescale of molecular vibrations of IR spectroscopy is of the order of 10^{13} s^{-1} which ideally complements those of ESR and NMR, 10^8 s^{-1} and 10^5 s^{-1}, respectively.

3. The principal advantage of FTIR compared with normal dispersive spectrometers is that spectra of higher signal-to-noise ratio in

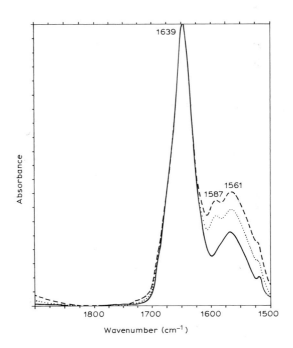

Fig. 7.4. Difference spectra of immunoglobin G in ²H₂O after subtraction of ²H₂O. Control immunoglobulin G (- - - -). Iron-stressed immunoglobin G at high radical-to-protein ratios, representative spectra (----, ...). IgG at 20 mg/ml is treated with a radical-generating system composed of iron(II) and hydrogen peroxide. Samples are placed in a thermostatically controlled Beckman FH-01 CFT microcell equipped with calcium fluoride windows and a 6 μm tin spacer. The spectrometer is continuously purged with dry air to avoid interference from water-vapour absorption. The spectra are recorded at 20°C by signal averaging 400 scans at a resolution of 4 cm⁻¹. Difference spectra were obtained by digital subtraction of buffer spectra from the corresponding sample spectra.

a given scanning time are obtained because the detector examines all of the scanned spectrum simultaneously (Fellgett's or the *multiplex advantage*) and the high optical throughput of the interferometer owing to its absence of slits (the Jacquinot advantage).

Disadvantages are that the single-beam FTIR usually has inferior ordinate stability compared with a double-beam dispersive instrument and the moving mirror is extremely sensitive to vibration.

7.3. Amino acid analysis for the identification of amino acid side-groups in proteins

Amino acid analysis is the general technique for detecting changes in the structure of individual amino acids on exposure of a protein to a variety of radical-generating systems.

7.3.1. Principle of the technique

Proteins are hydrolysed into their constituent amino acids and rendered positively charged in a low pH buffer. Amino acid separation and analysis is achieved by a cation exchange chromatography process. Separation is effected by ion exchange chromatography (Scheme 7.1) on a resin consisting of negatively charged, sulphonated beads of cross-linked polystyrene functioning as a cation exchanger:

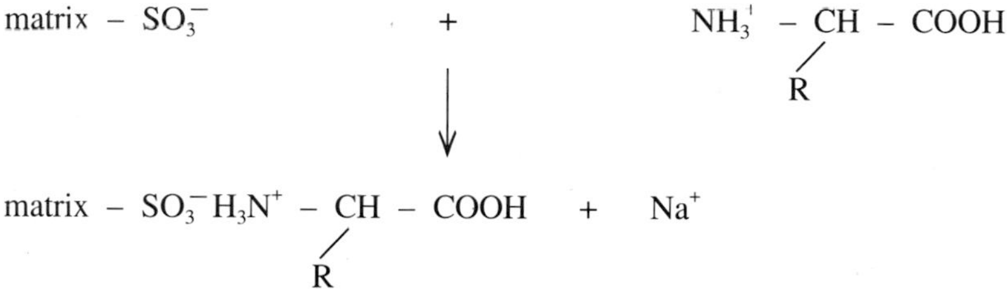

$$\text{matrix} - SO_3^- \qquad + \qquad NH_3^+ - CH - COOH$$
$$\diagup$$
$$R$$

$$\downarrow$$

$$\text{matrix} - SO_3^- H_3N^+ - CH - COOH \quad + \quad Na^+$$
$$\diagup$$
$$R$$

Scheme 7.1.

The conditions of the column are altered by gradually increasing the pH, the temperature and the concentration of the buffer counter ions. By gradually increasing the pH (to 3.2 to 4.25 to 6.45) the amino acids reach their isoelectric points, at which point the ionic attraction to the resin is lost and the amino acid elutes from the column. Amino acids of similar isoelectric points are separated by increasing the temperature ($50 \to 80°C$) and the counter-ion concentration (Na^+, $0.2 \to 2$

M), thus moving the equilibrium of the exchange process (depicted above) to the left.

7.3.2. Detection of amino acids

Two types of detection system are in common use with amino acid analysers, ninhydrin detection and fluorescence detection. The two systems differ in that fluorescence detection is more sensitive than ninhydrin detection, but it is more specific in that it does not detect amino acids such as proline. The detection reagent is mixed with the eluate from the column and the mixture passes into the fluorimeter or spectrophotometer. The system described here is based on the ninhydrin reaction with the separated amino acids.

7.3.3. Procedure

7.3.3.1. Preparation of solutions
Analar grades of all chemicals must be used and all buffers must be filtered through a 0.22 μm filtration membrane before use to remove particulate material.
1. 10 mM sodium chloride in Nanopure water.
2. 1.25 mM norleucine
3. Hydrolysing mix (to be prepared immediately prior to use):
   10 ml 6 N HCl (1:1 dilution of stock Aristar HCl with nanopure water)
   100 μl 10% phenol (10 g phenol made up to 100 ml in 6 N HCl)
   4 μl 2-mercaptoethanol
   0.1 g EDTA (i.e., 0.1%)
4. Loading buffer:
   0.2 M sodium citrate, pH 2.2
5. Separation buffers:
   0.2 M sodium citrate, pH 3.2, 4.25, 6.45 (2 M)
6. Ninhydrin reagent:
   ethylene glycol (ethanediol), 1575 ml
   ninhydrin powder, 22.5 g

sodium/potassium acetate buffer, 675 ml
hydrindantin, 1.8 g
To prepare sodium/potassium acetate buffer:
potassium acetate, 647.68 g
sodium acetate trihydrate, 299.20 g
tripotassium citrate, 8.8 g
glacial acetic acid, 220 ml
deionized water to 2.2 litres

All solid chemicals should be dissolved in approximately 1.5 litres of deionized water before the addition of the acetic acid. The volume is made up to 2.2 litres in a measuring cylinder, and a sample taken for pH determination. On dilution 1:3 with water a pH 5.20 ± 0.03 should result.

To prepare the ninhydrin reagent

The ethylene glycol is poured into a ninhydrin reservoir bottle which has been soaked in RBS (a phosphate-free alkaline surface-active cleaning agent) and washed thoroughly. Nitrogen is bubbled through slowly for approximately 15 min. The ninhydrin is added and allowed to dissolve (45 min) with gentle stirring and constant nitrogen flow. A portion of the acetate buffer is filtered (Whatman GF/A paper) and 675 ml added to the reservoir bottle. After mixing for 15 min the hydrindantin is added. The stirring, with nitrogen purging, is continued for a further 15 min and the reservoir bottle loaded into the analyser. The reagent should be left for 24 h before analysis is attempted.

7.3.3.2. Preparation of samples for hydrolysis

1. Before the samples are hydrolysed for amino acid analysis, they must be rendered salt-free. Samples are thus dialysed twice against 2 litres of 10 mM sodium chloride in Nanopure water or against a volatile buffer such as 0.5% ammonium bicarbonate.
2. Samples are then transferred to new 13×100 mm glass culture tubes that have been washed in RBS.
3. Samples should contain about 50 to 100 μg protein, equivalent to at least 2 nmol of the least abundant amino acid.

10 μl of 1.25 mM norleucine are added to act as an internal standard. After sealing the tubes with parafilm, the samples are frozen in a bath made from solid CO_2 carefully added to ethanol; after piercing the parafilm a few times with a needle, the samples are left to lyophilize overnight (for volumes of 1 ml and above).

Small samples may be dried in the liquid state, a procedure which tends to get rid of contaminating ammonia better than lyophilization but runs the risk of sample loss by bumping. Samples can be placed in a desiccator containing a Petri dish of P_2O_5 (*care* with water) and evacuated overnight to dry.

7.3.3.3. Hydrolysis of samples

500 μl of the hydrolysis mixture is added to the dried samples in the tube. Using a very hot flame, the middle of the tube is heated and subsequently pulled out to yield a neck approximately 1.5 inches long (long enough to be easily sealed yet short enough to prevent breakage on vortexing). The tubes are connected to the vacuum pump and evacuated while vortexing. The neck of the tube is then sealed (if possible putting a small handle on the sealed tube).

The sealed hydrolysis tubes are transferred to the oven, equilibrated to 110°C (for 4 h prior to insertion of samples), for 24 h. After this period, the samples are removed from the oven, cooled, and at this stage can be stored at −20°C if they cannot be processed immediately.

The tubes are centrifuged briefly on a bench centrifuge. The tubes are opened by making a scratch round the top of the tube with a diamond pencil, cracking by touching the scratch with a hot piece of glass rod and the top knocked off. The samples are dried by transferring the tubes to a desiccator containing sodium hydroxide and evacuated for at least 1 h, using a sodium hydroxide trap to protect the pump. The samples are left under vacuum overnight in the sealed desiccator. The dried hydrolysates can be stored at −20°C, if necessary.

7.3.3.4. The amino acid analyser

The basic schematic structure of the amino acid analyser is shown in

Fig. 7.5. The sample containing the mixture of amino acids from the hydrolysed protein is stored in the autoloader and loaded onto a column of cation-exchange resin.

Buffers at varying ionic strength and pH values are pumped through the column to separate the various amino acids.

The column eluate is mixed with ninhydrin reagent and the mixture passed through the high-temperature reaction coil where the amino acid-ninhydrin complex is formed. (The amount of the coloured ninhydrin complex formed is directly proportional to the quantity of amino acid present in the eluate.)

The mixture thereafter is fed into the photometer unit where the amount of light absorbed at specific wavelengths (see below) indicates the amount of each amino acid-ninhydrin complex.

The photometer output is connected to a two channel chart recorder, one channel for the 570 nm output, the other channel for the 440

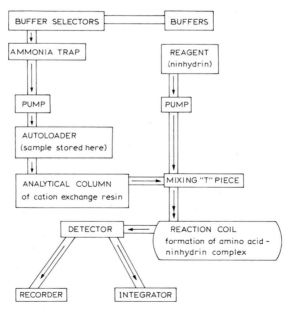

Fig. 7.5. Simplified layout of the amino acid analyser.

nm output. The amino acid concentrations are recorded as a series of peaks.

7.3.3.5. Separation of hydrolysed amino acids

The sample is resuspended in 200 μl of loading buffer (pH 2.2). An aliquot of 20–100 μl is loaded into a loading capsule and the sample injected onto the top of the column.

As the amino acids come off the column they are accessed by the ninhydrin reagent. The ninhydrin-amino acid complexes (Schemes 7.3 and 7.4) are detected spectrophotometrically at 570 nm (primary amino acids) and 440 nm (secondary amino acids, such as proline) and recorded as a series of peaks (Fig. 7.6). The retention time of the peak on the chart identifies the amino acid, the area under the peak indicating the quantity of the amino acid present.

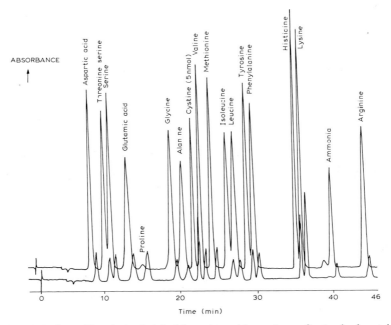

Fig. 7.6. Amino acid analysis of ninhydrin-amino acid complexes of a standard protein hydrolysate. 10 nmol of each amino acid was applied, except where stated otherwise.

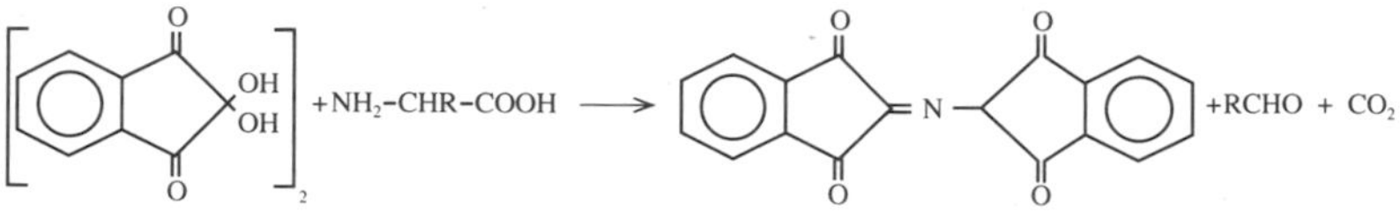

Scheme 7.2. Simplified reaction between ninhydrin and amino acids

A calibration analysis is performed prior to the sample analyses to produce a standard trace for comparison purposes.

7.3.4. Precautions

1. If 'ready-to-use' buffers are not to be used, then buffers prepared in the laboratory must conform to the same criteria of purity and performance:
   only salts of the highest purity should be used;
   when preparing buffers the salts should be dissolved in distilled and freshly deionized water – the purity of the water has a serious effect on the quality of the buffer. Salts should also be dried before use.
2. The column temperature must be accurately controlled, this being varied as necessary to produce the required separation.
3. The major outside source of contamination to buffers is the ingress of ammonia from the atmosphere. Glassware may also be a source of contamination: all glassware for use with amino acid analysis

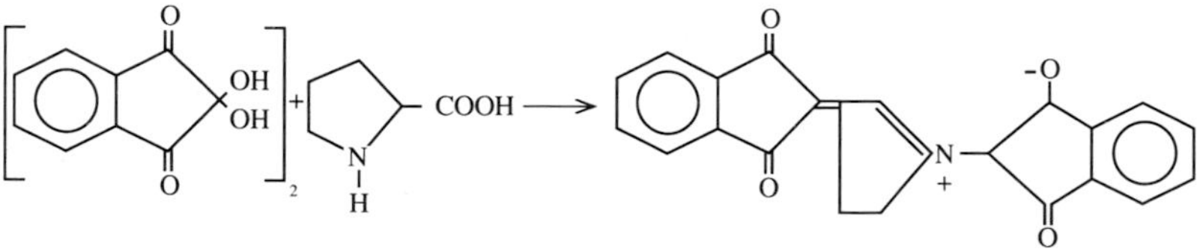

Scheme 7.3. Simplified reaction between ninhydrin and imino acids

should be rinsed well with distilled, deionized water to remove any dust or surface film.

4. Handling glassware may also introduce contaminants: gloves should be worn to prevent contamination with amino acids from the skin and untouched pipette tips should be used throughout.

7.4. Fluorescamine assay for the availability of side-chain amino groups

Proteins adjacent to peroxidizing phospholipids may undergo modification by the interaction of the free amino groups of the side-chains of the substituent lysine residues with a range of aldehydic metabolites of lipid hydroperoxide breakdown.

Alternatively, free amino groups of lysine side-chains may be oxidized. Thus the fluorescamine assay is used to measure the extent of loss of amino groups in proteins.

7.4.1. Principle

Fluorescamine reacts with primary amines to form a fluorescent product (Scheme 7.4). The fluorescence of a solution containing protein with fluorescamine is directly proportional to the number of free amino groups.

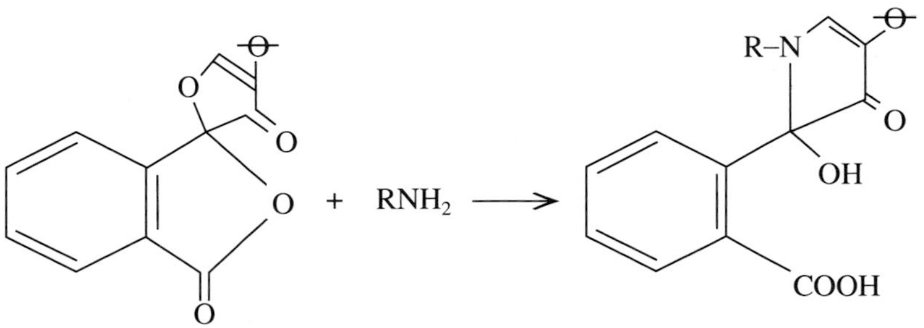

Scheme 7.4.

7.4.2. The procedure

The following solutions are prepared (Udenfried et al., 1972):
1. 0.03 g of fluorescamine in 100 ml of dioxane.
   N.B. Plastic boats should be avoided as the fluorescamine solution dissolves plastic.
2. 0.05 M phosphate buffer, pH 8.
3. A standard solution of albumin (bovine serum) of concentration 4 mg/ml. Define the concentration accurately by diluting in water to give A_{279} of 0.140, corresponding to a concentration of 200 μg/ ml (determined from the molar absorption coefficient at 279 nm for albumin in aqueous solution of 45 000). Dilute this solution 1:4 to give a final concentration of 50 μg/ml.

Standard incubations are set up in duplicate in the range of 0–10 μg albumin (0–200 μl of prepared standard) in 5 mM phosphate buffer (up to 3 ml volume).

The pH of each incubation is checked and adjusted if necessary to pH 8–8.5.

The various samples to be investigated, containing an amount of protein in the range of the assay in 3 ml, are set up in triplicate.

To each incubation 1 ml of the fluorescamine solution is added rapidly using a glass syringe whilst vortexing the solution. The fluorescence of each sample is measured between 5–30 min after the addition of the fluorescamine at an excitation wavelength of 390 nm, scanning the fluorescence over the emission wavelength range of 440–500 nm.

From the standard plot, the number of free amino groups in the sample under investigation can be determined.

7.4.3. Precautions

Fluorescamine interacts with water very rapidly to give non-fluorescent hydrolysis products. Thus, glassware should be thoroughly dry.

7.5. Detection and quantification of thiol side-chains

Radicals can affect thiol groups in proteins and peptides in a number of different ways. The response might be formation or loss of disulphide bridges. More dramatic free radical exposure can lead to the formation of sulphenic, sulphinic and sulphonic acids. The loss of free thiol groups can be assessed utilizing the methods described here.

7.5.1. Introduction and principle

Conversion of thiols to disulphides is a useful marker of oxidative events. Thiol groups are important in stabilizing and maintaining the functional tertiary structure of many proteins including membrane proteins. Thiol groups also have a potential role in biological redox couplings (Chance, 1981; Meister and Anderson, 1983), in oxygen metabolism (Sies and Akerboom, 1984), in radiation protection (Bump et al., 1982), and the sensitivity of cells to heat and thermal tolerance (Mitchell and Russo, 1983), amongst others. Thus, measurement of thiol redox status in cells and membranes is an effective marker of membrane and protein oxidation.

RSH groups are not easy to assay. Some arise during the handling of the biological material before the measurements are made and some arise from the lack of specificity of many reagents used as probes for quantitation of sulphydryl compounds (reviewed in Russo and Bump, 1989).

Spectroscopic methods of detection and quantitation of thiol groups have dominated the analytical procedures. An ideal reagent for thiol group determination would have the following properties:
1. selectivity
2. large molar absorption
3. a chromophore whose absorbance differs from those of normal biological chromophores
4. highly and stoichiometrically reactive to thiols and stable in the reaction solution over a large pH range
5. water soluble

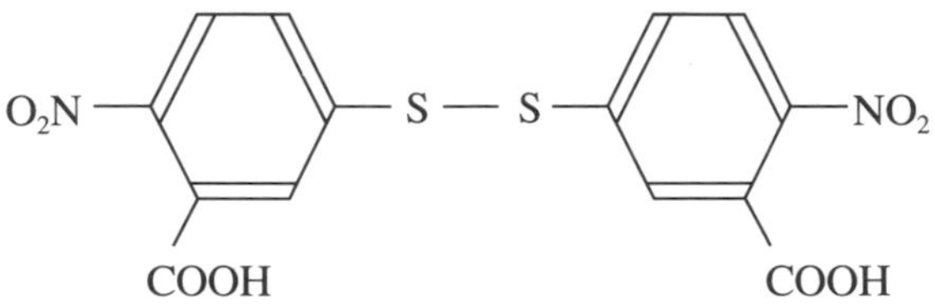

Scheme 7.5.

6. no interfering chromogenic properties until it has reacted with the sulphydryl groups
7. react equally with the protein and non-protein thiol groups
8. small enough to allow access to hidden protein thiol groups within non-denatured proteins.

Probably the most frequently used spectrophotometric method to detect thiol groups, both for non-protein and protein sulphydryl groups involves the use of Ellman's reagent (Scheme 7.5).

5,5'-Dithiobis(2-nitrobenzoic acid), (DTNB) (Ellman 1959) undergoes disulphide exchange with thiol groups and the formation of 5-thio-2-nitrobenzoate anion (TNB) (Scheme 7.6).

The thioquinone has an absorption maximum at 412 nm with $E = 13\,600$ $M^{-1} \cdot cm^{-1}$ which can be quantified spectrophotometrically. The reagent is specific for thiol groups, is applied at pH 8.0 and is

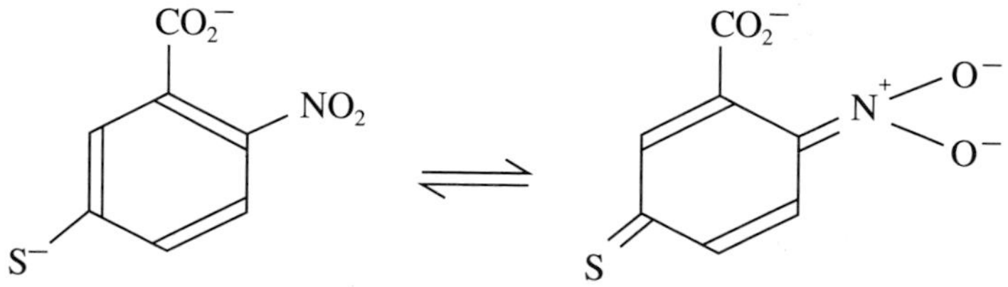

Scheme 7.6.

rapid. Ideally the measurement of TNB should be carried out between pH 8 and 9, the lower the pH the less the absorbance of the chromophore. DTNB can also be used to measure disulphides. Thiol groups have to be blocked with such reagents as N-ethylmaleimide or iodoacetate, and then RSSR groups are reduced to RSH groups. Released TNB can then be measured (Habeeb, 1966). By following the rate of production of TNB, Ellman's reagent can be used to evaluate the accessibility of the free thiol groups in proteins (Haest et al., 1978). Thus low final rates indicate the presence of inaccessible SH groups. Presumably some may completely avoid interaction.

7.5.2. Procedure

7.5.2.1. Preparation of solutions
1. 5,5'-Dithiobis(2-nitrobenzoic acid) (mM) (DTNB) (prepare freshly prior to use). 0.0198 g made up to 50 ml phosphate buffer, pH 8.0.
2. Sodium dodecyl sulphate (10%) (SDS)
   5 g made up to 50 ml in 5 mM phosphate buffer, pH 8.0.
3. Glutathione stock solution (prepare freshly just prior to use) 50 mg in 25 ml 5 mM phosphate buffer, pH 8.0. Dilute 10-fold in 5 mM phosphate buffer, pH 8.0, to 0.2 mg/ml or 0.651 μmol/ml.

7.5.2.2. Protocol
1. To 0.3 ml of protein/membranes of known protein concentration, 0.3 ml SDS are added and the solution mixed thoroughly.
2. 2.4 ml phosphate buffer are then added and the solutions mixed.
3. The background absorbance at 412 nm is then measured.
4. 0.3 ml of the thiol reagent DTNB is then added, the solution mixed and incubated for 1 h at 37°C.
5. The absorbance is measured at 412 nm.
6. It is also preferential to run blanks with 0.3 ml membranes, 0.3 ml SDS, 2.4 ml 5 mM phosphate buffer, pH 8.0, instead of DTNB.

Standard curve:
7. A range of reduced glutathione standards in 0.3 ml of phosphate

buffer are prepared (usually 0–195 nmol) from a stock solution (taking 0–0.30 ml stock).

8. Proceed as above without reading the absorbance at 412 nm before the colour development.

Using this method the value for fresh normal human erythrocyte membranes is about 80 nmol/mg membrane protein.

7.5.2.3. Problems

TNB absorbs at 412 nm and haem-containing proteins also absorb strongly in that region. Hence the TNB method is unsuitable for systems containing haem proteins, such as haemoglobin or myoglobin. To avoid such interference, 4,4'-dithiopyridine is used in place of TNB.

(The resulting 2-thiopyridone has an absorption maximum at 343 nm, $E = 8080$ $M^{-1} \cdot cm^{-1}$. The magnitude of the absorbance is independent of pH from 1 to 8. Thus the reaction with sulphydryl groups can be followed spectrophotometrically.)

7.6. Adaptation of the method for estimating the thiol content of membrane proteins

When analysing the thiol content of membrane proteins, underestimates may arise due to the inaccessibility of some of the SH groups within the hydrophobic portion of the membrane. The following method is used for accurate assessment of membrane protein thiol groups (Deuticke et al., 1988)

7.6.1. Solutions

1. 1% (w/v) sodium dodecyl sulphate in 5 mM phosphate buffer, pH 8.0.
2. 3 mM 4,4'-dithiopyridine
   33.05 mg in 50 ml water.

Moderate heating and vigorous stirring are required to dissolve completely.

3. Stock reduced glutathione

   115.24 mg glutathione dissolved in 100 ml isotonic phosphate buffer, diluted 10-fold in phosphate buffer to give 115.24 μg/ml or 375 nmol/ml.

7.6.2. Procedure

1. 0.4 ml of membranes of known protein concentrations are added to 2.4 ml SDS in phosphate buffer and mixed thoroughly.
2. The absorbance is measured both at 324 nm and 406 nm to carry out the correction for the membrane-associated haemoglobin determination, the absorbance at the latter wavelength to be subtracted.
3. 0.14 ml of 4,4' dithiopyridine solution is added.
4. The samples are incubated at 37°C for exactly 15 min, and subsequently incubated at room temperature for exactly 15 min.
5. The absorbance is then measured at 324 nm.
6. Blanks are run with:

   0.4 ml membrane

   2.4 ml SDS in phosphate buffer

   0.14 ml distilled water (in place of the colour reagent).

   This will measure the endogenous A_{324} and avoid the necessity of measuring the background absorbance before adding the diothiopyridine reagent.

Standard curve

7. This is prepared from a range of reduced glutathione standards in 0.4 ml volume, ranging from 0–150 nmol (0 – 0.4 ml stock solution) made up to 0.4 ml with phosphate buffer.
8. The procedure is continued as above from the addition of SDS, in step 1.

Or alternatively, assessments can be made by accurate correlation with the absorption coefficient $E = 22\,000$ M^{-1} · cm^{-1} at 324 nm.

7.7. Fluorescence detection of aromatic amino acid side-chains

7.7.1. Introduction

It is the aromatic amino acid side-chains that give proteins their fluorescent characteristics. If these aromatic groups are damaged, the fluorescence is modified. However, phenylalanine fluorescence is not observed in the presence of other aromatic groups and tyrosine fluorescence is detected only in the absence of tryptophan (Teale, 1960). Even in proteins containing high relative proportions of tyrosine to tryptophan, the fluorescence of the former is masked.

Spectra

Proteins are assessed for oxidative modification by measuring the following:

1. the reduction in the emission intensity of the intrinsic fluorescence of tryptophan; intrinsic tryptophan fluorescence on excitation in the wavelength range 279–298 nm occurs in the range 320–350 nm, depending on the protein being studied. Shifts in the emission maximum of the intrinsic fluorescence are an indication of the modified environment of the side-chain, reduction in the fluorescence intensity indicates a structural modification.

2. the appearance of N-formylkynurenine, an oxidative breakdown product of tryptophan;

    Induced fluorescence λ_{ex} 360 λ_{em} 454 nm in neutral or acid solution is indicative of N-formylkynurenine, an oxidative breakdown product of tryptophan. The sensitivity of this assay can be enhanced by altering the pH to 10.5 with borate buffer and measuring the fluorescence characteristics at 400 nm, after excitation at 315 nm, the modified fluorescent characteristics being accompanied by a 20-fold enhancement in intensity under these conditions (Pirie, 1972; Gutteridge and Wilkins, 1983).

3. the appearance of modified tyrosine residues.

    The appearance of bityrosine can be assessed on excitation at 325 nm and observing the fluorescence at 410–420 nm (Prutz et al., 1983).

7.7.2. Calibration of the spectrofluorimeter

The spectrofluorimeter is calibrated using 0.1 μg/ml quinine sulphate in 0.1 M sulphuric acid.

With slits set at 10 and the lowest sensitivity setting the fluorescence of quinine sulphate is excited at 350 nm and the emission measured at 445 nm.

7.7.3. Cautionary points

Protein concentration should be < 0.5 mg/ml to prevent excessive absorption interfering with the fluorescence emission intensity. The samples should be optically clear with no turbidity.

7.8. Protein carbonyl content as a marker of protein oxidation

7.8.1. Procedure

This assay is based on the interaction between carbonyl groups, resulting from free radical-modified proteins and 2,4-dinitrophenyl-hydrazine (see Fig. 5.9) (Oliver et al., 1987). As well as isolated proteins, protein in cells can also be studied by this method, as described below.

(For the preparation of carbonyl-free solvents see Section 5.3.2.2.) The cells are washed three times in serum-free medium and harvested by scraping. The harvested cells are washed again with PBS, pH 7.4. Cell disruption is effected by light sonication and after centrifugation (600 × g, 15 min) the insoluble material is removed. The supernatant fraction is retained and the protein concentration assessed by the Lowry assay.

The supernatant fraction of the cell preparations or the protein solutions to be studied are divided into two equal aliquots each containing in the region of 1 mg protein; one is assessed for protein carbonyl content with dinitrophenylhydrazine, the other is the control. Both aliquots are precipitated with 10% (w/v) TCA (final concentration) and the supernatant discarded. One sample is suspended in 2 N HCl and the other with an equal volume of 0.2% (w/v) dinitrophenylhydrazine (DNPH) in 2 N HCl.

The samples are incubated at 37°C in small 10 ml conical flasks in a shaking water bath for 1 h. The samples are then precipitated with 10% trichloroacetic acid (final concentration) and subsequently extracted with ethanol/ethyl acetate (1:1, v/v), discarding the latter layer. Samples are then reprecipitated with 10% trichloroacetic acid (final concentration) and the supernatants again removed.

The pellets are carefully drained and dissolved in 6 M guanidinium hydrochloride in 20 mM sodium phosphate buffer, pH 6.5. Solutions are centrifuged at $6000 \times g$ at 4°C, the supernatant carefully removed and then re-centrifuged to obtain an optically clear solution.

The difference spectrum of the DNPH-treated sample versus the acid-treated control is determined and the results can be expressed as nmol DNPH incorporated/mg of protein, based on average absorptivity of $21.0 \text{ mM}^{-1} \cdot \text{cm}^{-1}$ in the region of 365–375 nm for most aliphatic hydrazones (Jones et al., 1956).

The absorbance of the DNPH-derivative of oxidatively modified albumin (1 mg/ml) treated with 100 μM Cu^{2+}/200 μM H_2O_2 for 30 min, above that of the control, is of the order of 0.07.

7.9. Chromatography and mass spectrometry (SIM); applied to study side-chain damage

This is an extremely powerful technique for detecting trace modifications to biological structures such as proteins and DNA bases (Karan and Simic, 1988). Typically, the damaged biopolymers are hydrolysed and the products dried and treated with bis(trimethylsilyl)trifluoroacetamide (BSTFA) to convert amino acids or breakdown products into more volatile trimethylsilyl derivatives. The products are separated, for example, on a fused-silica capillary column and detected using a mass spectrometer that is tuned to a specific mass (selective ion monitoring: SIM). The advantage of this method is that other fragment ions of the selected mass that would normally interfere are removed because the retention time on the gas chromatograph will differ from that of the product of interest.

In studying protein damage by hydroxyl radicals, possibly the best product for analysis is (2-hydroxyphenyl)alanine. Other hydroxylated products are, of course, formed, but this is unique to hydroxyl radical attack, and can readily be separated from other products.

CHAPTER 8

Determination of radical damage to DNA

8.1. Introduction

This chapter describes broadly some methods which can be applied to investigate how radicals can damage DNA in intact cells and isolated DNA molecules.

In general radical damage to DNA has not been extensively studied except by photochemical and radiolysis methods. By far the most extensive studies are those involving radiation damage, and these studies typify radical damage in general.

Detailed description of the wealth of techniques available for studying such systems is beyond the scope of this book. However, the authors have attempted to provide a broad overview of some of the approaches which can be undertaken for identification of radical damage which DNA may have experienced in intact cellular systems or for standard assessment of the ways in which DNA may be damaged by these reactive species. As well as the use of cell systems we also use liquid-phase studies on dilute solutions of DNA to illustrate the effects of ˙OH radical damage, and solid-state studies to show the effects of electron-gain and -loss.

8.2. Radiation effects on dilute solutions of DNA

It seems to be generally agreed that the hydroxyl radical is the main source of damage, and we confine our attention to this radical. Its reactions are sometimes maximized using N_2O to convert solvated electrons into more ˙OH radicals.

In principle, one might expect ˙OH radicals to attack at phosphate, deoxyribose or DNA base sites. The results do not require attack on phosphate, but do require both hydrogen transfer from C-H units of the sugar and addition to all the bases. Current opinion seems to be that the latter reactions are of comparable importance, although opinion *varies*. Hydrogen transfer, in free competition with base addition, would be a lot slower, and therefore make only a minor contribution. However, it seems that for duplex DNA (but less so for single-stranded DNA) there is some form of steric or entropy control that favours C-H attack, so that both processes are of great importance. This results in two alternative modes for strand-breakage, one rapid (following H-transfer from the sugar) and one relatively slow (following base addition).

8.2.1. Strand breaks

Most studies of DNA damage centre on strand breaks, partly because they are relatively easy to study. Also, there is a large (but controversial) body of opinion that implicates double strand breaks (DSB) as the major lesion leading to cell death. It seems that single strand breakage (SSB) and base damage that does not lead to strand breakage are less important.

There are a variety of techniques available for studying strand breakage. These are in general use in all studies of DNA and are in no sense special for radical studies. We therefore confine ourselves to a brief summary of the methods.

8.2.1.1. Electrophoretic methods

(a) Use of small supercoiled DNAs   It is now clear that most of the DNA from animal and bacterial sources is restrained in the cell in a supercoiled or superhelical configuration. Small viral and plasmid DNA can readily be isolated in this supercoiled form (Dabis et al., 1986). One approach to detecting both SSBs and DSBs caused by the direct effect of radical species on DNA, is to analyse their effects on

the migratory properties of small supercoiled DNAs in agarose under the influence of an electric field.

Plasmids (e.g., pBR322), viral DNA (e.g., SV40 DNA) or the replicative form of OX174 bacteriophage (see Ueda et al., 1985) occur as covalently closed supercoiled cyclic DNA molecules. As such they have no breaks in either strand. This supercoiled state (form I) is intrinsically less stable than the uncoiled state and the breakage of a single strand instantly converts a *supercoiled* cyclic DNA molecule into its simple *relaxed* cyclic state (form II). Because cyclic DNA molecules become increasingly compacted as they become more supercoiled, 'relaxed' DNA is easily distinguished from supercoiled DNA by its slower sedimentation in a centrifugal field and by the longer time it takes to move through an agarose gel under the influence of an electric field. Thus the extent of single-strand breakage can be followed by analysing electrophoretic behaviour of small supercoiled DNA molecules in agarose gels 0.5 – 2% (see Boffey, 1983) before and after oxidant treatment, the faster migrating supercoiled form I being converted to the slower moving relaxed form II (see Fig. 8.1). A single double-strand break on the other hand gives a linear DNA molecule (form III) which migrates in agarose gels at a position intermediary between form I and form II (see Fig. 8.1).

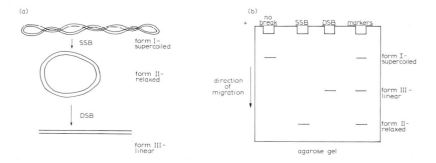

Fig. 8.1. **(a)** DNA forms: form I, supercoiled; form II, relaxed; form III, linear. **(b)** Schematic representation of agarose gel electrophoresis of forms I, II and III.

(b) Use of linear DNA Agarose gel electrophoresis in its normal form is not useful for separating long linear strands of DNA, since they move as a broad unresolved band. The technique of pulsed-field electrophoresis overcomes this problem, since it involves periodic molecular reorientation, and the longer the DNA strand, the longer it takes to reorient. In the Transverse Alternating Field Electrophoresis (TAFE) system, the gel is held vertically with both sides exposed to electrolyte, and the electrodes are placed on each side with alternating field passing through the gel slab horizontally. The DNA strands oscillate with the field but migrate in parallel lanes. Very large DNA strands can be separated using this technique (Gardiner et al., 1986).

Comparisons between this technique and neutral elution (see below) for detecting DSBs in irradiated DNA show good agreement, which is satisfying. Probably the TAFE system will become the method of choice.

Protocol
0.3 μg of supercoiled DNA in 20 μl of buffer (pH 7.4) are exposed to a source of radical species. The reaction is terminated by adding 5 μl of 0.1 M EDTA containing 50% sucrose and 0.1% bromophenol blue (to act as marker for clectrophoresis) and the mixture subjected to electrophoresis through agarose gels (0.5 to 2%) (see Ueda et al., 1985).

Watch-points
1. A limitation is that multiple SSBs will only give form II, unless two breaks occur close together in alternate strands, giving a DSB and hence form III.
2. It is important to use a minimum of a non-reactive buffer such as phosphate, with no radical scavenging properties.

8.2.1.2. Elution methods
These can be used on cellular DNA isolated, for example, from cells previously exposed to oxidant or other stress. Methods in major use involve using tritiated thymidine labelling or fluorescence tags. The former method is described here.

Alkaline elution techniques   In one type of method the cellular DNA is denatured in alkali followed by hydroxyapatite chromatography which separates single stranded DNA from double stranded DNA (Britten and Kohne, 1965). The production of single-stranded DNA is related to the number of SSBs in the DNA which act as unwinding points. As an example the effects of UV-irradiation on the integrity of mammalian cell DNA has been studied using this approach (Collins, 1977). Initially mammalian cells in culture are extensively labelled with [³H]thymidine (see Adams, this series, 1980).

METHOD

1. Such cells are subjected to UV-irradiation in growth medium at around 5×10^5 ml^{-1}.
2. After treatment they are washed with cold saline, and the cells are resuspended in cold distilled water at $(2-4) \times 10^5$ cells in 0.1 ml and pipetted onto the surface of 1 ml alkaline lysis solution (5% sucrose containing 0.3 M NaOH, 0.5 M NaCl) in a glass vessel (2 cm i.d.) cooled on ice.
3. After lysis (approx 5 min) the solution is neutralized with 0.52 ml 1 M KH_2PO_4 and mixed vigorously.
4. 1 ml 1% (w/v) sodium dodecyl sulphate is then added, and the lysate warmed until the SDS dissolves.
5. The lysate is then passed through a 25 G needle to shear the DNA (the precise number of passages does not appear to be too critical).
6. After dilution to 20 ml with distilled water lysates can be stored overnight at 4°C before hydroxyapatite chromatography.
7. Hydroxyapatite (DNA-grade, Bio-Rad Laboratories) is suspended in 1 mM sodium phosphate, pH 6.8, and poured to a height of 0.5 cm in a 0.9 cm diameter water-jacketed column maintained at 70°C. Each lysate should be adsorbed onto a fresh hydroxyapatite column and the initial eluate collected and the retained ³H-labelled DNA eluted using prewarmed 70°C phosphate buffer, pH 6.8, of increasing molarity, i.e., 0.1 M (2×2 ml), 0.2 M (2×2 ml), 0.3 M (1×4 ml) and finally 4 ml 0.5 M NaOH.
8. To assess the level of [³H]DNA in each fraction, the DNA can be

precipitated using the ice-cold 5% trichloroacetic acid (TCA) onto
2.5 cm Whatman GF/C glass-fibre filters. After washing with 5%
TCA and 96% ethanol, the filters are dried and the ^{3}H-radioactivity
determined by liquid scintillation spectrometry.

9. The level of radioactivity in a single-stranded DNA is taken as an
   indication of SSBs. Calculation of the approximate number of
   breaks requires reference to the molecular size of the cellular ge-
   nome and comparison with the known number of breaks induced
   in genomes of specific sizes by X-ray and UV-irradiation of known
   dosage (see Collins, 1977).

Neutral elution techniques   To detect DSBs it is possible to make use
of 'neutral elution' techniques. Such neutral elution techniques have
been described by Bradley and Kohn (1979) and by Mullinger and
Johnson (1985).

METHOD

1. As was the case for the alkaline elution technique described above,
   the DNA of the cultured cell has to be first extensively labelled
   with [^{3}H]thymidine (see Adams, this series, 1980).
2. After exposure of cells to radiation or oxidant treatment, between
   $(2-4) \times 10^5$ cells are loaded at 40°C onto a Nucleopore 25 mm, 2
   μm pore size, polycarbonate filter (Bio-Rad Laboratories) on a
   filter support.
3. The cells are lysed on the filter at pH 9.6 at room temperature in
   the solution containing 0.05 M Tris, 0.05 M glycine, 0.025 M Na$_2$-
   EDTA, 2% (w/v) sodium dodecyl sulphate containing 0.5 mg/ml
   freshly dissolved *proteinase K* (Boehringer, Ltd.). This is a broad
   specificity bacterial protease which will work in SDS.
4. Elution is carried out over a period of 15 h using the same buffer
   minus the proteinase K. Ten '90 min' fractions are collected. The
   ratio of radioactivity eluted to that remaining bound to the filter
   is taken as an indicator of the DSBs in the sample (see Mullinger
   and Johnson, 1985).

8.2.1.3. Sedimentation methods

Use of nucleoids   A very sensitive method for detecting DNA damage within mammalian cells has been described by Cook and Brazell (1976). When mammalian cells are gently lysed in the presence of non-ionic detergents and high salt concentrations, structures resembling nuclei are released. These are called 'nucleoids' and contain nearly all nuclear RNA and DNA but are depleted of nuclear proteins. Their DNA is compact so that the nucleoids sediment more rapidly in sucrose gradients than those from cells whose DNA contains single-strand breaks.

METHOD

In the case of human white blood cells, 50 μl of a suspension containing $(2–5) \times 10^5$ cells in phosphate-buffered saline (PBS) is layered onto 150 μl of lysis mixture floating on top of 4.6 ml 15–30 sucrose gradients containing 1.95 M NaCl, 0.01 M Tris, 0.001 M EDTA in addition to various concentrations of ethidium bromide which binds to DNA of different extents of supercoiling to simplify the differential density characteristics necessary for separation by this technique. The lysis mixture should contain sodium chloride, EDTA, and Triton X-100 in amounts which on addition of 1 vol of cells in PBS to 3 vol of the lysis mixture will give final concentrations of 1.95 M NaCl, 0.1 M EDTA and 0.5% Triton X-100. Fifteen min after the addition of the cells to the lysis mixture the gradients are spun at 30 000 rpm for 25 min at 20°C in the SW50.1 Beckman rotor. The gradient is fractionated. The position of the nucleoids can be determined by measurement of absorbance at 254 nm of each fraction. The distance travelled by the nucleoids from experimentally treated cells is expressed as a ratio relative to that travelled by nucleoids from control cells sedimenting under the same conditions but in the presence of ethidium bromide. As the concentration of ethidium bromide in the sucrose gradients is increased the distance travelled by the nucleoids falls to a minimum and then rises again (see Cook and Brazell, 1976). The introduction of any single strand breaks during experimental treatment of the cells, will reduce the sedimentation of the nucleoids and abolishes the biphasic response to EB which is characteristic of super-coiled DNAs. It can be concluded therefore that nucleoid DNA is su-percoiled and that the supercoils are lost by the introduction of single stranded breaks. Agents that break DNA can be detected in 'nuc-leoids' with the sensitivity, rapidity and economy usually associated with bacteria.

Use of alkaline sucrose gradients   Another but less sensitive means of detecting SSBs in mammalian cell DNA is the use of alkaline sucrose density gradients (see Meneghini, 1976). As was the case in the use of the elution methods (Section 8.2.1.2) an essential preliminary step is the extensive labelling of cellular DNA by incubation of cell cul-tures with [^{3}H]thymidine (Adams, this series, 1980). These labelled cell cultures can then be variously exposed to potential DNA-damaging agents.

METHOD
Following such treatment, the cultured cells in monolayer are washed with phosphate-buffered saline and extracted in 4 ml 0.5% Triton X-100 in saline/EDTA (100 mM NaCl, 10 mM EDTA, pH 8.0) for 2 min at room temperature. This releases most of the cytoplasmic mate-rial whilst the nuclei remain attached to the culture dish. 0.5% sodium dodecyl sulphate and 40 μg/ml pancreatic RNAse (preincubated at 80°C for 10 min, to inactivate DNAse) in saline/EDTA (above) is then added and the mixture incubated for 20 min at 37°C. One vol-ume of chloroform/isoamyl alcohol (20:5, v/v) is then added and the phases mixed gently. The aqueous phase is separated by centrifuga-tion and extracted again with chloroform/isoamyl alcohol. DNA is precipitated from the aqueous phase with 2 vol 95% ethanol and resus-pended in 0.01 M Tris, pH 7.5. Alkaline sucrose density gradients (5–20%) are prepared in 0.1 M NaCl, 0.1 M NaOH with a final volume of 4.1 ml. Samples of DNA (max 3 μg) are layered on the top of these gradients and spun at 32 000 rpm at 20°C in a SW.50.1 Beckman rotor for 120 min. Fractions are collected and the [^{3}H]DNA precipitated

onto Whatman GF/C glass fibre discs (2.5 cm) using ice-cold 5% tri-chloroacetic acid. After washing discs with further 5% trichloroacetic acid, ethanol and acetone, the level of ^{3}H-radioactivity is determined by scintillation spectrometry. T2 bacteriophage DNA and λ-bacterio-phage DNA can be used as molecular weight markers in such alkaline sucrose gradients (their molecular weights being taken as 67×10^6 and 16.3×10^6, respectively) and the molecular weight corresponding to each fraction can be determined according to Studier (1965).

8.2.1.4. Other methods
In pulse radiolysis or flash photolysis studies, light scattering or con-ductivity methods are used to estimate strand breaks (see Section 8.3.3). These studies show that there are two rate processes involved. The fast process is assigned to ˙OH radical attack on C-H bonds of the sugar with rapid β-elimination to give strand breaks. The slow process starts with ˙OH radical addition to one of the bases. This is followed at some stage by H-atom transfer from a nearby C-H bond of a sugar unit, again leading to strand breakage. This process is easy to understand for the pyrimidines which, after ˙OH addition to C_6, have an active radical centre able to undergo H-atom transfer. It is less easily understood for the purines and may not occur in the case of guanine attack.

8.2.2. Base damage detected by gas chromatography-mass spectrometry with selected ion monitoring

Again, in pulse radiolysis studies, attempts are made to detect base-damage radical centres optically. This is not so easy for duplex DNA, because there are many possible structures and optical spectra are rather broad and featureless.

Considerable success is now being achieved using analytical meth-ods after exposure. The DNA is broken down into monomeric units which are first separated chromatographically, then analysed with a mass spectrometer with selected ion monitoring (SIM).

For example, Dizdaroglu and Bergtold (1986) used a GC-MS-SIM apparatus (Chapter 7) to study hydroxyl radical and H· atom attack on dilute aqueous DNA and were able to quantify a range of damaged bases in the 10 fmol range, with doses in the 0.1 to 10 Gy range.

The products of hydroxyl radical-mediated DNA scission (Dizdaroglu, 1986; Von Sonntag, 1987) include (Fig. 8.2):

5,6-dihydrothymine,
5-hydroxy-5,6-dihydrothymine,
5-hydroxymethyluracil,
5-hydroxyuracil,
5-hydroxycytosine,

5,6-dihydrothymine   5-hydroxy-5,6-dihydrothymine   8-hydroxyadenine

5-hydroxymethyluracil   5-hydroxyuracil   5-hydroxycytosine

thymine glycol   cytosine glycol   8-hydroxyguanine

4,6-diamino-5-formamidopyrimidine   2,6-diamino-4-hydroxy-formamidopyrimidine

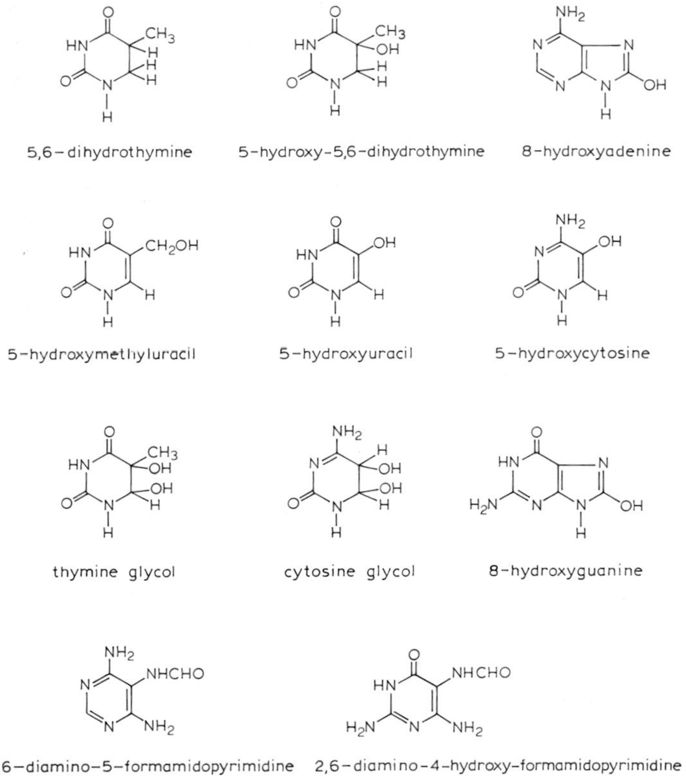

Fig. 8.2. The chemical nature of products of hydroxyl radical attack on the constituent bases of DNA.

5,6-dihydroxycytosine,
thymine glycol,
cytosine glycol,
4,6-diamino-5-formamidopyrimidine,
2,6-diamino-4-hydroxy-5-formamidopyrimidine,
8-hydroxyadenine,
8-hydroxyguanine.

Clearly hydroxyl radical addition is fairly indiscriminate, as expected. This is a very powerful method for analysing base damage not only in radical chemistry, but in any reactions involving base damage. The problem with base damage, as with side-group damage in proteins, is that very low concentrations of the products of damaged bases are involved, and need to be detected in the presence of large concentrations of undamaged material.

One of the best techniques is the combination of gas-chromatography and selective ion monitoring using a mass spectrometer. This is briefly described in Section 7.8. It depends on prior knowledge of the types of modified bases that are expected and their ionization patterns in the mass spectrometer.

Hence radical damage to DNA can be detected by the characterization and quantitative measurement of the products derived from the purine and pyrimidine bases. The method is demonstrated here by applying model systems of DNA treated with superoxide-generating systems or hydrogen peroxide in the presence of an iron chelate (Aruoma et al., 1989). Products of radical-mediated base damage in DNA are measured using GC-MS-SIM after acid hydrolysis of DNA and trimethylsilylation.

See Fig. 8.3 for chromatogram showing base modifications obtained after exposure of DNA to hydroxyl radicals (for this example, radiolytically induced).

PROTOCOL
Total volume 1.2 ml containing:
solution of DNA (calf thymus) at a final concentration 0.5 mg/ml
    hypoxanthine 0.33 mM;

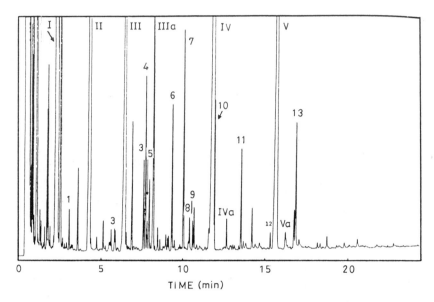

Fig. 8.3. An example of a gas chromatogram of the base-damage to DNA after exposure to hydroxyl radicals (generated radiolytically) – study of the trimethyl-silylated acidic hydrolysate of the modified DNA (modified from Dizdaroglu, 1988; with permission). *Peaks*: **I**, phosphoric acid; **II**, thymine; **III** and **IIIa**, cytosine; **IV** and **IVa** adenine; **V** and **Va** guanine; **1** uracil, **2** 5,6-dihydrothymine; **3** 5-hydroxy-5,6-dihydrothymine; **4** 5-hydroxyuracil; **5** 5-hydroxy-5,6-dihydrouracil; **6** 5-hydroxycytosine; **7** *cis*-thymine glycol; **8** *trans*-thymine glycol; **9** 5,6-dihydroxyuracil; **10** 4,6-diamino-5-formamidopyrimidine; **11** 8-hydroxyadenine; **12** 2,6-diamino-4-hydroxy-5-formamidopyrimidine; **13** 8-hydroxyguanine.

100 μM iron(II) chloride or iron(III) chloride plus 120 μM EDTA;
10 mM KH_2PO_4/KOH buffer, pH 7.4;
xanthine oxidase 10.08 units/ml diluted immediately prior to use.

The reaction is started by adding the xanthine oxidase and the mixture incubated for 1 h at 37°C. Solutions are then dialysed against water at 4°C for 3 days.

In order to determine the concentration of DNA the absorbance is measured at 260 nm.

For a detailed account of the practical aspects of this technique the reader is referred to Dizdaroglu (1985, 1986). The hydrolysis and derivatization procedures are clearly described in Aruoma et al. (1989).

8.3. Direct radiation damage to DNA

In DNA, guanine bases are good electron donors, and the pyrimidine bases are fairly good electron acceptors. Since the bases are stacked in the DNA duplex, with fair overlap between the π-orbitals, some co-operation is expected to occur between bases to facilitate electron gain or loss reactions. These can be induced chemically, but have largely been studied by radiolysis either of 'dry' DNA, or of frozen aqueous solutions.

ESR results suggest that such direct damage is remarkably simple, leading, in the first stage, to just two major trapped centres, electron-capture at thymine ($T^{\cdot-}$) or cytosine ($C^{\cdot-}$) and electron-loss at guanine ($G^{\cdot+}$). It seems that electrons or holes generated locally on solvating water, phosphate or sugar units, all migrate rapidly. The positive and negative centres ($G^{\cdot+}$ and $T^{\cdot-}/C^{\cdot-}$) are generally well separated from each other at the time of initial trapping, and almost certainly become 'fixed' by proton transfer. There is some controversy regarding the nature of the primary anion-radical, which is thought by some to be $C^{\cdot-}$ rather than $T^{\cdot-}$. However, it is reasonably certain that both radical anion centres are involved. $T^{\cdot-}$ centres are efficiently converted into a carbon-protonated derivative, TH [8.1] in the region of 200K. The fate of the $C^{\cdot-}$ centre is less clear.

[8.1]

Concomitant strand-break studies on frozen aqueous systems on irradiation suggest that both radical cations and radical anions can lead to strand breaks. They also show that DSBs are remarkably frequent under these conditions. This was interpreted in terms of proximal trapping of positive and negative radical centres, each leading to a strand break (Boon et al., 1984).

The conclusions are that for DNA tightly packaged as chromatin in cell nuclei, direct interaction with radiation will give these radical ions. Furthermore, electrons ejected from water close to the chromatin will often add to DNA before becoming solvated, and also hole centres ($H_2O^{\bullet+}$) can react with water to give electron-transfer [8.2] as:

$$H_2O^{\bullet+} + H_2O \rightarrow H_2O + H_2O^{\bullet+} \qquad [8.2]$$

an alternative to proton loss, and hence can also migrate to DNA augmenting the damage. Thus the effective target for damage is the DNA-histone complex together with 'solvating' water.

The radical-ions can lead to SSBs and to DSBs, but also, of course, to base damage. Preliminary results with chromatin suggest that hole centres are readily trapped within the proteins, but some electrons can pass into the DNA strands to give $T^{\bullet-}$ and $^{\bullet}TH$ radicals (Cullis et al., 1987).

8.4. Damage at the quaternary level

It has been known for some time that radical agents and especially very low doses of ionizing radiation can greatly alter the density of packed DNA. In our previous discussion attention has been confined to specific DNA lesions, but nuclear function requires the correct organization of chromatin, and disruption at the quaternary level could be important. There are probably many routes to such damage, but here we confine discussion to one route which it is suggested should have a high quantum yield, and might lead to momentary or permanent damage.

8.4.1. The role of copper ions in the quaternary structure of DNA

As well as being coiled around the histone proteins, DNA is bound at intervals to nuclear structural protein. At least some of these links are thought to occur via Cu(II) ions, though the precise mode of bonding is, as yet, unknown. The situation envisaged by Cramp and his co-workers (George et al., 1987) is that electron capture at Cu(II) could give Cu(I) ions. These tend to bind ligands less strongly than Cu(II), so the protein-DNA linkage may break. More subtly, whilst Cu(II) ions tend to adopt a square-planar configuration, Cu(I) prefers to be tetrahedral. Such a change must seriously modify the chromosomal organization in the vicinity of the copper. Both changes help to explain the marked decrease in density for DNA-protein suspensions.

On exposure to ionizing radiation, a major and sometimes neglected product is hydrogen peroxide. This reactive molecule has a relatively long life, and should often survive long enough to react with Cu(I) [8.3].

$$Cu(I) + H_2O_2 \rightarrow Cu(II) + \, ^{\cdot}OH + OH^- \qquad [8.3]$$

These regenerate the copper(II), but the hydroxyl radicals, generated close to DNA, have a high probability of attacking it to give, say, an SSB. A repeat of this process could give a DSB. The key factor is the very high electron scavenging ability of Cu(II) ions.

8.5. Free radicals and DNA structural analysis

For a variety of purposes, procedures have been described to ascertain the nature of the complexes between regulatory proteins and DNA. An early approach was to form a complex between the DNA and the protein and then to subject this complex to digestion with deoxyribonuclease I. The particular region of DNA in specific contact with the protein is protected by the latter from enzymic digestion (see Kadonaga et al., 1986). Another approach has been to analyse the protection afforded by specific proteins to specific DNA sequences from cleavage by dimethyl sulphate (Gidoni et al., 1985). This reagent normally cleaves unprotected DNA at guanine residues.

These 'footprinting' analyses, based on enzymic and chemical digestion, are now widely used to define DNA (and RNA) and their complexes with various ligands. Recently active radical probes have been used as footprinting agents in protection assays in a variety of systems (e.g., Tullius and Dombroski, 1986; Chalepakis and Beato, 1989; Hayes and Tullius, 1989; Schickor et al., 1990). Such probes rely on active radical intermediates, most likely hydroxyl radicals, released by Fe(II) in the presence of an electron donor, probably via a Fenton reaction. In addition, hydroxyl radicals also appear to react with DNA in a conformation-specific manner which may allow some prediction of DNA secondary structure (see Burkhoft and Tullius, 1987; Zorbas et al., 1989; Lu et al., 1990).

Appendix A

1. Chemicals, columns and accessories

albumin (bovine serum)
(Sigma Chemical Co.)
anasyl H (Analabs Inc.)
L-ascorbic acid (Merck)
bleomycin sulphate
(Sigma Chemical Co.)
t-butyl hydroperoxide (Aldrich Chemical
Co.)
butylated hydroxytoluene (2,6-di-tert-
butyl-p-cresol) (Sigma Chemical Co.)
caeruloplasmin (human)
(Sigma Chemical Co.)
catalase (Sigma Chemical Co.)
Chelex resin (Bio-Rad)
cholesterol esterase
(Sigma Chemical Co.)
cyanmethaemoglobin kit (Boehringer)
1,3-cyclohexanedione
(Aldrich Chemical Co.)
cytochrome C (Sigma Chemical Co.)
cytochrome C peroxidase (Sigma
Chemical Co.)
decenal (Sigma Chemical Co.)
deoxyribose (Sigma Chemical Co.)
desferrioxamine mesylate (CIBA-Geigy)
deuterium oxide (Sigma Chemical Co.)
dimethyl sulphoxide (Merck)
5,5'-dimethyl-1-pyrroline-N-oxide
(DMPO) (Aldrich Chemical Co.)
3-(4,5-dimethylthiazol-2-yl)-2,5-diphenyl
tetrazolium bromide (see MTT)
2,3-dihydroxybenzoic acid
(Sigma Chemical Co.)
2,5-dihydroxybenzoic acid
(Sigma Chemical Co.)
2,4-dinitrophenylhydrazine
(Sigma Chemical Co.)
5,5'-dithiobis(2-nitrobenzoic acid)
(DTNB) (Sigma Chemical Co.)

4,4'-dithiopyridine (Sigma Chemical Co.)
DNA-calf thymus (Sigma Chemical Co.,
Pharmacia)
DNA-super-coiled (Pharmacia)
5,8,11,14-[1-^{14}C] eicosatetraenoic acid
(Amersham)
β-eleostearic acid (Alltech Associates,
U.K.)
ferrozine {3-(2-pyridyl)-5,6-bis
(4-phenylsulphonic acid)-1,2,4-
triazine} (Sigma Chemical Co.)
fluorescamine {4-phenylspiro[furan-
[2-^{3}H]phthalan]-3,3'-dione} (Sigma
Chemical Co.)
glutathione (reduced)
(Sigma Chemical Co.)
hepes (Merck)
hexanal (Sigma Chemical Co.)
horseradish peroxidase
(Sigma Chemical Co.)

HPLC columns and accessories:
Altex Analytic columns (Hichrom)
Carbohydrate Analysis Column
(Waters Associates)
Celite 545 columns (for preparation of
carbonyl-free solvents) (Merck)
Chromosorb stainless steel column
(Varian Associates)
ERC column (Emma Optical
Works Ltd.)
Exsil (Hichrom)
Extrelut Column (Merck)
HPLC Stainless steel tubing and valves
(Pierce Chemical Co.)
Lichrosorb columns (Applied
Chromatography Systems)
Octadecyl Silicon Gel Column (JT
Baker)
Sep-Pak sample preparation cartridges
(Jones Chromatography, Waters
Associates)

Spherisorb (Hichrom)
Spherogel TSK (Anachem)
Waters Reversed Phase Columns (Waters
    Associates)
Zorbax Analytical Columns (Hichrom)
HPLC grade solvents: (Rathburn's
    Chemicals)
methanol, acetonitrile, acetic acid,
    dichloromethane, propan-2-ol
hydrindantin (Sigma Chemical Co.)
hydrogen peroxide (Merck)
hydroxyapatite (Bio-Rad Laboratories)
4-hydroxynonenal (gift from Professor
    H. Esterbauer, University of Graz)
hypoxanthine (Sigma Chemical Co.)
immunoglobulin G (Sigma
    Chemical Co.)
lipoxygenase (Soybean) (Sigma
    Chemical Co.)
lucigenin (5-amino-2,3-dihydro-1,4-
    phthlazinedione) (Boehringer)
luminol (10,10'-dimethyl-9,9'-
    bisacridinium dinitrate) (Boehringer)
malonaldehyde bisacetal (Sigma
    Chemical Co.)
mannitol (Sigma Chemical Co.)
MTT (Sigma Chemical Co.)
myoglobin (horse heart) (Sigma
    Chemical Co.)
NBT (Sigma Chemical Co.)
ninhydrin (Sigma Chemical Co.)
nitric acid (70% spectrosil) (Merck)
Nucleopore polycarbonate filter (Bio-
    Rad Laboratories)
oxalic acid (Sigma Chemical Co.)
Percoll (Pharmacia)
α-phenyl-t-butylnitrone (PBN) (Aldrich
    Chemical Co.)
phenylmethylsulphonyl fluoride (PMSF)
    (Sigma Chemical Co.)
phorbol myristate acetate (Sigma
    Chemical Co.)
phospholipase A_2 (*Naja naja* venom)
    (Sigma Chemical Co.)
proteinase K (Boehringer)
quinine sulphate (Merck)

RBS – phosphate-free alkaline surface-
    active cleaning agent (Chemical
    Concentrates)
salicylic acid (Aldrich Chemical Co.)
scopoletin (Aldrich Chemical Co.)
selenium (elemental) (Aldrich
    Chemical Co.)
Sephadex G-15 (Sigma Chemical Co.)
silicic acid column (Merck)
sodium diatrizoate (Sigma Chemical Co.)
superoxide dismutase (Sigma Chemical
    Co.)
2-thiobarbituric acid (Sigma
    Chemical Co.)
thiourea (Sigma Chemical Co.)
trichloroacetic acid (Merck)
water – sterile entotoxin-free (Travenol)
water – nanopure (Rathburn's
    Chemicals)
xanthine (Sigma Chemical Co.)
xanthine oxidase (Sigma Chemical Co.)

2 Major equipment suppliers

1. EPR spectrometers:
   (endor, spin-echo)
   Bruker Ltd., Coventry, Warwickshire
   Jeol (U.K.) Ltd, Welwyn Garden
   City, Herts AL7 ILT, U.K.

2. Atomic absorption spectrometers
   Phillips PU 9200, U.K.

3. HPLC equipment
   Waters Associates, Harrow,
   Middlesex
   Beckman (System Gold) Instruments
   (U.K.) Ltd.
   High Wycombe, Bucks HP12 4JL,
   U.K.

4. UV/VISIBLE spectrophotometers
Beckman DU 65, DU 70, diode array
model 168
High Wycombe, Bucks HP12 4JL,
U.K.

5. Spectrofluorimeters
Perkin-Elmer Ltd, Beaconsfield,
Bucks HP9 1QA
Hitachi Scientific Instruments,
Finchampstead, Wokingham, Berks
RG11 4QQ, U.K.

6. Fourier transform infra-red
spectrometers
Perkin-Elmer Ltd., Beaconsfield,
Bucks HP9 1QA, U.K.

7. Gel Electrophoresis - agarose
Bio-Rad Laboratories Ltd., Hemel
Hempstead, Herts HP2 7TD, U.K.

8. Liquid scintillation counters
Beckman (LS6000 range), High
Wycombe, Bucks HP12 4JL, U.K.

3. Chemicals suppliers' addresses

Aldrich Chemical Co., Gillingham, Kent
SP8 4JL, U.K.

Amersham International PLC,
Aylesbury, Bucks HP20 2TP, U.K.

Anachem, Luton, Bedfordshire, U.K.

Analabs Inc., New Haven, CT, U.S.A.

Applied Chromatography Systems,
Macclesfield, Cheshire, U.K.

J.T. Baker, Chillipsburg, NJ, U.S.A.

Bio-Rad Laboratories Ltd., Hemel
Hempstead, Herts HP2 7TD, U.K.

Boehringer-Mannheim U.K. Ltd.,
Lewes, East Sussex BN7 1LG, U.K.

Chemical Concentrates (RBS) Ltd.,
London SW18 1RE, U.K.

CIBA-Geigy, Horsham, Sussex, U.K.

Emma Optical Works Ltd., Tokyo,
Japan

Hichrom, Reading, Berkshire, U.K.

Jones Chromatography, Llanbradach,
Glamorgan, Wales, U.K.

Merck Ltd., Dagenham, Essex IRM8
1RF, U.K.

Pharmacia Ltd., Milton Keynes, Bucks
MK9 3HP, U.K.

Pierce & Warriner (U.K.) Ltd., Upper
Northgate Street, Chester CH1 4EF,
U.K.

Rathburn's Chemicals, Peebleshire,
Scotland, U.K.

Sigma Chemical Co., Poole, Dorset,
U.K.

[Travenol], now Baxter Health Care,
Newbury, Berks RG16 0QW, U.K.

Varian Associates, Walton-on-Thames,
Surrey KT12 2QF, U.K.

Waters Associates, Harrow, Middlesex,
U.K.

Appendix B – Methods for isolation of low density lipoproteins and erythrocyte membranes and cytoskeletons from blood

1. Blood manipulation

Anticoagulant: *Heparin*: Commercially supplied heparinized tubes (Vacutainers)

      or   *Acid citrate dextrose (ACD)*: Glucose (0.114 M), trisodium citrate (30 mM), sodium chloride (72.6 mM), citric acid (2.81 mM) pH 6.4

Blood samples: Blood is taken into heparin or ACD anticoagulant for isolation of erthrocyte membranes and into ACD anticoagulant for isolation of low density lipoproteins (LDL). All manipulations of blood should be initiated as soon as possible after venipuncture in the case of LDL preparation and within 17 h if preparing erythrocyte membranes. Blood samples are stored at 4°C before use.

2. LDL extraction and purification (see Mills et al. in this series)

Reference: Modification of Chung (1980) by G. Paganga (Division of Biochemistry, United Medical & Dental Schools of Guy's & St. Thomas's).

(**a**) For LDL preparation ACD is measured into a universal tube to a volume 1/5 that of the blood sample to be added.

(**b**) The universal tubes are placed in the IEC Centra-7R centrifuge, care being taken to use only a single black holder since a double holder will cause the universal tube to break. Centrifuge for 20 min at 3×1000 rpm on this centrifuge setting.

(c) The plasma layer is removed by syringe (without disturbing the red blood cells) enabling measurement of the volume of the plasma yield (remembering that part of the volume is 1/5 ACD). The plasma density is approximately 1.006 g/ml.

(d) The plasma needs to be a density of approximately 1.3 g/ml. So by using Radding & Steinberg formula it is possible to determine the weight of sodium bromide to be added.

$$M = v(d_2 - d_1)/1 - \bar{v}d_2$$

M is the weight of sodium bromide to be added to a volume v of solution to change its density from d_1 to d_2 (at a stated temperature); and $\bar{v}$ is the partial specific volume of the salt at the relevant temperature and concentration. Values of $\bar{v}$ can be determined for sodium bromide applying the data of Baxter and Wallace (see Mills et al.).

Simplified, plasma volume $\times$ 0.4571 gives the weight of NaBr required to give the correct density.

(e) The measured NaBr is dissolved into the plasma using a magnetic stirrer. It is advised not to use lumps of NaBr, stirring should be slow so as to avoid froth formation.

(f) The Beckman centrifuge tubes (3 ml vol.) are filled with 0.9% NaCl and 1 ml plasma layered under avoiding air bubbles. The tubes are then sealed.

(g) Centrifugation is carried out using an angled rotor (TLA 100.3) which must be capped with aluminium caps. The following parameters must be set on the Beckman 100 Ultracentrifuge; Time 20 min; Speed 100 K; Temperature 16°C; Rotor TLA 100.3. When the cycle is complete, vacuum is pressed to enable the door to be opened. Pliers may be needed to help pull the tubes from the rotor.

(h) The test tubes are placed in a suitable rack and the tube vacuum removed by inserting a needle at the top. The crude LDL is collected using a syringe and 1 ml transferred to a new Beckman centrifuge tube already containing 0.8 ml of 1.151 g/ml density solution. The tube is filled with density solution 1.063 g/ml, sealed as before and mixed.

(i) The tubes are centrifuged using the same parameters as before. This is the washing phase of the LDL.

(j) LDL is distributed at the top of the tube and is obtained by syringe as before. The LDL solution is placed into dialysis tubes presoaked in Tyrodes buffer and dialysed for 2 h, using a mechanical shaker and changing the buffer every 30 min.

(k) The dialysate is collected and concentrated to about 2 ml, using the following parameters on the Sorvall centrifuge: Temperature 4°C; time 45 min; speed 8000 rpm.
N.B. The tubes must be balanced using the buffer.

(l) The LDL is stored at 4°C.

3. Separation of erythrocytes from whole blood

Whole blood is centrifuged at 600 × g for 10 min at 4°C in graduated centrifuge tubes using a swing out rotor. The plasma and buffy coat (a layer of white blood cells over the erythrocytes) are removed by aspiration. Packed erythrocytes are washed twice in 10 vol. of isotonic 5 mM phosphate buffer pH 7.4, or other buffer as required, by centrifugation at 600 × g for 10 min at 4°C.

4. Density separation of erythrocyte sub-populations

Reference: Nash et al., 1989.

Solutions: *Hepes buffer:* Hepes (20 mM), sodium chloride (135 mM), potassium chloride (5 mM), magnesium chloride (1 mM), disodium hydrogen phosphate (1 mM), glucose (5 mM), pH 7.40, osmolarity 300 mOsm/kg H_2O.
Percoll-diatrizoate solution: 10.0 ml of sodium diatrizoate in water (50%, w/v) are mixed with 45.0 ml Percoll. Hepes buffer (10.0 ml) is added to 49.5 ml of the above mixture and the osmolarity is adjusted

to 300 mOsm/kg H_2O by adding water. The volume is then made up to 75.0 ml with Hepes buffer.

Procedure: Packed erythrocytes, washed twice in Hepes buffer, are re-suspended in an equal volume of the same buffer. The erythrocyte sus-pension (2 ml) is carefully layered over the Percoll-diatrizoate solution (2 ml) in a glass centrifuge tube with an internal diameter of 10 mm. After centrifugation at 400 × g for 20 min at room temperature, erythrocytes with normal density are carefully removed from the sur-face of the Percoll-diatrizoate cushion using a pasteur pipette. This fraction is referred to as the normal density fraction. Erythrocytes with elevated density, which had pelleted below the Percoll-dia-trizoate, can then be removed in a similar manner. This latter fraction is referred to as the dense fraction. Finally, each fraction of erythro-cytes is washed twice in 10 vol. of Hepes buffer at 4°C.

5. Preparation of erythrocyte membranes

Reference: Dodge et al., 1963.

Procedure: Packed, washed erythrocytes are lysed by adding 10 vol. of 5 mM phosphate buffer pH 7.4 (at 4°C) while mixing. After leaving on ice for 30 min, the erythrocyte membranes are packed by centrifu-gation at 20 000 × g for 10 min at 4°C (Sorvall SS-34 rotor) and the haemoglobin-containing supernatant is removed by aspiration. The erythrocyte membranes are then washed three times by resuspending in fresh buffer (same volume as used for lysis) followed by centrifuga-tion under the same conditions. Finally, the membranes are resus-pended in isotonic 5 mM buffer (same volume as used for lysis) fol-lowed by centrifugation under the same conditions. Finally, the membranes are resuspended in isotonic 5 mM phosphate buffer pH 7.4 and repacked. Erythrocyte membranes are quantified on the basis of protein concentration using the assay of Lowry et al. (1951) (see Section 7).

6. Isolation of the erythrocyte membrane cytoskeleton

Principle: In this procedure erythrocytes are treated with Triton X-100 which is reported to solubilize the membrane lipid leaving the underlying cytoskeletal network intact. The cyto-skeletons are separated from cytosolic components, Triton and solubilized lipid by centrifugation through a sucrose solution. The high salt concentration of the sucrose solution ensures the removal of residual lipid and integral membrane proteins from the cytoskeletal network.

Reference: Pinder and Gratzer 1983.

Solutions: *Triton X-100 extraction solution*: Triton X-100 (150 mg/ml), Hepes (24 mM), sodium chloride (0.15 M), disodium EDTA (0.5 mM), dithiothreitol (0.5 mM), pH 7.0 (prepared fresh each day)
Sucrose wash: Sucrose (30%, w/v), Hepes (24 mM), potassium chloride (0.6 M), ATP (0.5 mM), disodium EDTA (0.5 mM), dithiothreitol (0.5 mM), pH 7.0 (prepared fresh each day)
Phenylmethysulphonyl fluoride (PMSF): PMSF (5%, w/v) in methanol (prepared fresh each day)

Procedure: PMSF is added to packed washed erythrocytes to give a final concentration of 0.05% (w/v) followed by thorough mixing by inversion. After 5 min the cells are resuspended in an equal volume of isotonic 5 mM phosphate buffer pH 7.4. The resultant erythrocyte suspension is mixed with an equal volume of Triton X-100 extraction solution by inversion and the mixture is rapidly layered on top of 20 ml of sucrose wash in a centrifuge tube. Cytoskeletons are pelleted through the sucrose wash by centrifugation at 20 000 rpm in a 6 × 36 ml MSE swingout rotor for 1 h. After centrifugation the upper layer (red with haemoglobin) is aspirated and the walls of the centrifuge tube are washed with isotonic 5 mM phosphate buffer pH 7.4 to remove residual haemoglobin and Triton X-100 before the lower sucrose wash (colourless) is removed leaving a firm pellet of cytoskeletal material. The cytoskeletons are washed by resuspending in 10 ml isotonic 5 mM phosphate buffer, pH 7.4, and centrifuging at 31000 × g for 30 min (Sorvall SS-34 rotor) and finally resuspended with 0.8 ml of the same buffer. The entire procedure is carried out at 4°C. Cytoskeleton preparations are quantified on the basis of protein content. The assays of Lowry et al. (1951) and Bradford (1978) can be used to determine the protein concentration of the initial preparations.

7. Quantification of erythrocyte membranes and cytoskeletons. The Lowry protein assay

Reference: Lowry et al., 1951.

Solutions: Copper sulphate $5H_2O$ (1%, w/v)
Sodium potassium tartrate (2%, w/v)
sodium carbonate (2%, w/v), sodium hydroxide (0.1 M)
Folin and Ciocalteu's reagent diluted 1:1 with water immediately before use
Standard protein stock solution: bovine serum albumin (4 g/l) in water

Procedure: The alkaline copper reagent was prepared just prior to use by mixing 1 vol. of copper sulphate solution with 1 vol. of sodium potassium tartrate solution followed by 98 vol. of the sodium carbonate/sodium hydroxide solution. Alkaline copper solution (5 ml) is added to 100 μl of sample, containing between 20–100 μg of protein, diluted with 900 μl of water. After 10 min 0.5 ml of diluted Folin and Ciocalteus' reagent is added. The absorbance at 750 nm is read after 30 min and within 90 min from the final addition.

Standard curve: A series of dilutions of bovine serum albumin in water are prepared containing between 0 and 100 μg protein in 1 ml volume and the assay is performed as described above on each dilution. The exact concentration of the bovine serum albumin stock solution is determined using the molar absorption coefficient of 45000 for bovine serum albumin at 279 nm after diluting to approximately 0.2 mg/ml.

8. The Bradford protein assay

Reference: Bradford, 1976.

Solutions: *Protein reagent*: Coomassie brilliant blue G-250 (100 mg) is dissolved in 50 ml 95% ethanol. To this solution 100 ml 85% (w/v) phosphoric acid is added and the resulting solution is diluted to a final volume of 1 l and filtered with Whatman 1 paper.
Standard protein stock solution: bovine serum albumin (4 mg/ml) in water

Procedure: To 0.1 ml of cytoskeleton preparation containing between 10 to 100 μg protein is added 5 ml of the protein reagent. The absorbance at 595 nm is measured after 5 min and before 20 min against a reagent blank prepared from 0.1 ml isotonic 5 mM phosphate buffer, pH 7.4, and 5 ml of the protein reagent.

Standard curve: A series of dilutions of bovine serum albumin in isotonic 5 mM phosphate buffer, pH 7.4, are prepared containing between 0 and 100 μg protein in 0.1 ml volume and the assay is performed as described above on each dilution. The exact concentration of the bovine serum albumin stock solution is determined using the molar absorption coefficient of 45 000 for bovine serum albumin at 279 nm after diluting to approximately 0.2 g/l.

9. References

Bradford, M.M. (1976) A rapid and sensitive method for the quantitation of microgram quantities of protein utilising the principle of protein-dye binding. Anal. Biochem. *72*, 248–254.

Chung, B.H. (1980) Preparative and quantitative isolation of plasma lipoproteins: rapid single discontinuous density gradient ultracentrifugation in a vertical rotor.

Dodge, J.T., Mitchell, C. and Hanahan, D.L. (1963) The preparation and chemical characteristics of hemoglobin-free ghosts of human erythrocytes. Arch. Biochem. Biophys. *100*, 119–130.

Lowry, O.H., Rosebrough, N.J., Farr, A.L. and Randall, R.J. (1951) Protein measurement with the Folin phenol reagent. J. Biol. Chem. *193*, 265–275.

Mills, G.L., Lane, P.A. and Weech, P.K. (1984) A Guide-book to Lipoprotein Technique (Burdon, R.H. and Van Knippenburg, P.H., eds.), Elsevier, Amsterdam.

Nash, G.B., Boghossian, S., Parmar, J., Dormandy, J.A. and Bevan, D. (1989) Alteration of the mechanical properties of sickle cells by repetitive deoxygenation: role of calcium and the effects of calcium blockers. Br. J. Haematol. *72*, 260–264.

Pinder, J.C. and Gratzer, W.B. (1983) Structural and dynamic states of actin in the erythrocyte. J. Cell. Biol. *96*, 768–775.

References

Chapter 1

Badger, R.M., Wright, A.C. and Whittock, R.F. (1965) J. Chem. Phys. *43*, 4345–4350.
Barry, B.A. and Babcock, G.T. (1987) Proc. Natl. Acad. Sci. USA *84*, 7099–8007.
Boon, J., Olm, M.T. and Symons, M.C.R. (1988) J. Chem. Soc., Faraday Trans. I *84*, 3334–3345.
DeMaster, E.G., Raiji, L., Archer, S.L. and Weir, E.K. (1989) Biochem. Biophys. Res. Commun. 163, 527–533.
Ehrenberg, A. and Reichard, P. (1972) J. Biol. Chem. *247*, 3485–3492.
Grant, G. (1988) Chem. Biol. Interact. *65*, 157–162.
Griffiths, D.R., Robins, G.V., Seeley, N.J., Chandra, H. and Symons, M.C.R. (1982) Nature *300*, 435–438.
Kanofsky, J.R. (1983) J. Biol. Chem. *258*, 5991–5993.
Kanofsky, J.R. (1984) J. Am. Chem. Soc. *106*, 4277–4278.
Liu, Y.-C., Liu, Z.-L., Chen, P. and Wu, L. (1988a) Sci. Sinica (B) *31*, 1062–1072.
Liu, Y.-C., Liu, Z.-L. and Han, Z.-H. (1988b) Rev. Chem. Intern. *10*, 269–289.
Moncada, S., Palmer, R.M.J. and Higgs, E.A. (1988) Hypertension *12*, 365–372.
Niehaus, W.G. (1978) Bioinorg. Chem. *7*, 77.
Palmer, R.M.J., Ferridge, A.G. and Moncada, S. (1987) Nature *327*, 524.
Sahlin, M., Petersson, L., Graslund, A. and Ehrenberg, A. (1987) Biochem. *26*, 5541–5548.
San Filippo, J., Romano, L.J., Chern, C.I. and Valentine, J.S. (1976) J. Org. Chem. *41*, 586.
Scurlock, R.D. and Ogilby, P.R. (1987) J. Phys. Chem. *91*, 4599–4602.
Symons, M.C.R., Eastland, W. and Denny, L.R. (1980) J. Chem. Soc., Faraday Trans. I *76*, 1868–1874.
Symons, M.C.R. and Stephenson, J.M. (1981) J. Chem. Soc., Faraday Trans I *77*, 1579–1583.

Chapter 2

Ames, B.N., Cathcart, R., Schwiers, E. and Hochstein, P. (1981) Proc. Natl. Acad. Sci. USA *78*, 6858–6862.
Aruoma, O.I. and Halliwell, B. (1987) Biochem. J. *241*, 273–278.
Babior, B.M. (1978) New Engl. J. Med. *298*, 659–668.
Bielski, B.H.J. (1985) J. Phys. Chem. Ref. Data *14*, 1041–1100.
Blandenaur, M.J. (1973) Introduction to Chemical Ultrasonics. Academic Press, London.
Briemer, L.H. (1988) Br. J. Cancer *57*, 6–18.
Bruckdorfer, K.R. (1989) Prost. Leukot. Essent Fatty Acids *38*, 247–254.

Burdon, R.H. and Rice-Evans, C. (1989) Free Rad Res. Commun. 6, 345–358.
Burton, G.W. and Ingold, K.U. (1984) Science 224, 569–573.
Burton, G.W. and Ingold, K.U. (1986) Acc. Chem. Res. 19, 194–201.
Butler, J., Hoey, B.M. and Lea, J.S. (1988) In: Free Radicals, Methodology and Concepts (Rice-Evans, C. and Halliwell, B., eds.), Richelieu Press, London, pp. 457–479.
Chandra, H. and Symons, M.C.R. (1987) Nature 328, 833–834.
Cohen, G. and Hochstein, P. (1963) Biochemistry 2, 1420–1428.
Davies, K.J.A. (1987) J. Free Rad. Biol. Med. 2, 155–173.
Davies, K.J.A. and Goldberg, A.L. (1987) J. Biol. Chem. 262, 8220–8226.
Davies, M.J. and Slater, T.F. (1987) Biochem. J. 245, 167–173.
Deby, C. and Deby-Dupont, G. (1980) In: Biological and Chemical Aspects of Superoxide and Superoxide Dismutase (Bannister, W.H. and Bannister, J.V., eds.), Elsevier, Amsterdam, pp. 84–97.
Di Maschio, P., Kaiser, S. and Sies, H. (1989) Arch. Biochem. Biophys. 274, 532–538.
Diplock, A.T. (1983) CIBA Found. Symp. 101, 45–53.
Esterbauer, H. (1985) In: Free Radicals in Liver Injury (Poli, G., Cheeseman, K., Dianzani, M.U. and Slater, T., eds.), IRL Press, Oxford, pp. 29–47.
Esterbauer, H. (1988) In: Free Radicals, Methodology and Concepts (Rice-Evans, C. and Halliwell, B., eds.), Richelieu Press. pp. 243–268.
Esterbauer, H., Streigl, G., Puhl, H. and Rothender, M. (1989) Free Rad. Res. Commun. 6, 67–75.
Frei, B., Stocker, R. and Ames, B. (1988) Proc. Natl. Acad. Sci. USA 85, 9748–9752.
Fridovich, I. (1983) Annu. Rev. Pharmacol. Toxicol. 23, 239–257.
Gutteridge, J.M.C. (1987) Biochim. Biophys. Acta 917, 219–223.
Gutteridge and Stocks (1981) CRC Crit. Rev. Clin. Lab. Sci. 14, 287–329.
Halliwell, B. (1988) Biochem. Pharmacol. 37, 569–571.
Halliwell, B. (1990) Free Rad. Res. Commun. 9, 1–32.
Halliwell, B. and Gutteridge, J.M.C. (1984) Biochem. J. 219, 1–14.
Halliwell, B. and Gutteridge, J.M.C. (1985) Mol. Aspects Med. 8, 89–193.
Hemler, M.E. and Lands, W.E.M. (1980) J. Biol. Chem. 255, 6253–6261.
Kuehl, F.A., Humes, J.L., Egan, R.W., Han, E.A., Beveridge, G.C. and Van Armer, C.G. (1977) Nature 265, 120–173.
Labeque, R. and Marnett, L. (1988) Biochemistry 27, 7060–7070.
Lands, W.E.M. (1979) Annu. Rev. Physiol. 41, 633–652.
McKay, P.B. (1985) Annu. Rev. Nutr. 5, 323–340.
Marcillat, D., Han, Y., Lin, S.W. and Davies, K.J.A. (1988) Biochem. J. 254, 677–683.
Miyamoto, T., Ogino, N., Yamamoto, S. and Hayaishi, O. (1976) J. Biol. Chem. 251, 2629–2636.
O'Brien, P.J. (1969) Can. J. Biochem. 47, 485–492.
Peterson, D.A., Gerrard, J.M., Rao, G.H.R. and White, J.G. (1981) Prog. Lipid Res. 20, 299–301.
Reisz, P., Berdahl, D. and Chisman, C.L. (1985) Environ. Hlth. Perspect. 64, 233.
Roschupkin, D.I., Talitsky, V.V. and Pelenitsyn, A.B. (1979) Photochem. Photobiol. 30, 635–643.
Sies, H. (1986) Angewandte Chemie (Int. Edn. Engl.) 25, 1058–1071.

Singh, A., Koroll, G.W. and Cundall, R.B. (1982) Radiat. Phys. Chem. *19*, 137.
Slater, T.F. (1984) Biochem. J. *222*, 1–15.
Steinberg, D., Parthasarathy, S., Carew, T.E., Khoo, J.C. and Witztum, J.L. (1989) New Engl. J. Med. *320*, 915–924.
Stocker, R., Glazer, A.N. and Ames, B.N. (1987) Proc. Natl. Acad. Sci. USA *84*, 5918–5922.
Symons, M.C.R. (1988) Free Rad. Res. Commun. *5*, 131–139.
Thornalley, P.J. and Vasak, M. (1985) Biochim. Biophys. Acta *827*, 36–44.
Tappel, A.L. (1980) In: Free Radicals in Biology, Vol IV (Pryor, W.A. Ed.) Academic Press, New York, pp. 2–47.
Tappel, A.L. and Dillard, C.J. (1981) Fed. Proc. *40*, 174–178.
Van der Ouderra, F.J., Buyenek, M., Ugteren, D.H. and Van Dorp, D.A. (1977) Biochim. Biophys. Acta *487*, 315–331.
Vile, G. and Winterbourn, C.C. (1988) FEBS Lett. *238*, 356–357.
Wolff, S.P. and Dean, R.T. (1986) Biochem. J. *234*, 399–403.
Wolff, S.P., Garner, A. and Dean, R.P. (1986) Trends Biochem. Sci. *11*, 27–31.
Yamamoto, S., Ohki, S., Ogino, N., Shimizu, T., Yoshimoto, T., Watanabe, K. and Hayaishi, O. (1980) Adv. Prostagland. Thromboxane Res. *6*, 27–34.

Chapter 3

Abragam, A. and Bleaney, B. (1973) Electron Paramagnetic Resonance of Transition Metal Ions, Oxford Press, London.
Albano, E., Lott, K.A.K., Slater, T.F., Stier, P., Symons, M.C.R. and Tomasi, A. (1982) Biochem. J. *204*, 593.
Alger, R.S. (1968) Electron Paramagnetic Resonance. Wiley Interscience, New York.
Allen, R.C. (1981) In: Bioluminescence and Chemiluminescence, Basic Chemistry and Analytical Applications (DeLuca, M.A. and McElroy, W.D., eds.), Academic Press, New York, pp. 63.
Allen, R.C. and Loose, L.D. (1976) Biochem. Biophys. Res. Commun. *69*, 245.
Allen, R.C., Stjernholm, R.L. and Steele, R.H. (1972) Biochem. Biophys. Res. Commun. *47*, 679.
Ambruso, D.R. and Johnston, R.B. (1981) J. Clin. Invest. *67*, 352–360.
Andreae, W.A. (1955) Nature *175*, 859–860.
Angerhofer, A., Massotti, R.J. and Bowman, K. (1988) Israel J. Chem. *28*, 227.
Asmus, K.-D., Janata (1982) Nato Adv. Study Inst. Ser C. 91–128.
Atherton, N.M. (1973) Electron Spin Resonance. Halsted Press, London.
Bannister, J.V., Bellavite, P., Davoli, A., Thornalley, P.J. and Rossi, F. (1982) FEBS Lett. *150*, 300–302.
Beck, G. (1979) Rev. Sci. Instrum. *50*, 1147.
Boveris, A., Cadenas, E. and Chance, B. (1980) Photobiochem. Photobiophys. *1*, 175–182.
Boveris, A., Cadenas, E., Reiter, R., Filipkowski, M., Nakase, Y. and Chance, B. (1980) Proc. Natl. Acad. Sci. USA *77*, 347.

Boveris, A. and Chance, B. (1973) Biochem. J. *134*, 707–716.
Boveris, A., Oshino, N. and Chance, B. (1972) Biochem. J. *128*, 617–630.
Britigan, B.E., Rosen, G.M., Chai, Y. and Cohen, M.S. (1986) J. Biol. Chem. *261*, 4426–4431.
Brivati, J., Stephens, A. and Symons, M.C.R. (1990) J. Magn. Res., in press.
Cadenas, E., Boveris, A. and Chance, B. (1980) Biochem, J. *187*, 131.
Cadenas, E. and Sies, H. (1984) Methods. Enzymol. *105*, 221.
Change, B., Sies, H. and Boveris, A. (1979) Physiol. Rev. *59*, 527–605.
Connor, H.D., Thurnman, R.G., Galizi, M.D. and Mason, R.P. (1986) J. Biol. Chem. *261*, 4542.
Crawen, P.A., Pfanstiel F. and DeRubertis, F.R. (1986) J. Clin. Invest. *77*, 850.
Eiben, K. and Fessenden, R.W. (1968) J. Phys. Chem. *89*, 925.
Eggleton, P., Crawford, N. and Fisher, D. (1989) In: Separations Using Aqueous Phase Systems, Applications in Cell Biology and Biotechnology (Fisher, D. and Sutherland, I.A., eds.) Academic Press, New York. pp. 137–144.
Fessenden, R.W. and Schuler, R.H. (1971) J. Chem. Phys. *39*, 2147–2156.
Fischer, S.M. and Adams, L.M. (1985) Cancer Res. *45*, 3130.
Fletcher, M.P. and Gasson, J.C. (1988) Blood *71*, 652–658.
Fujimoto, S., Ishimitsu, S., Kanazawa, H., Mizutani, T., Ohara, A. and Haayakawa, T. (1987) Agric. Biol. Chem. *51*, 2851–2853.
Green, M.R., Hill, H.A.O., Okolow-Zubkowska, J. and Segal, A.W. (1970) FEBS Lett. *100*, 23–26.
Green, T.R., Fellman, J.H. and Eicher, A.L. (1985) FEBS Lett. *192*, 33–36.
Greenwald, R.A., Rush, S.W., Moak, S.A. and Weitz, Z. (1989) Free Rad. Biol. Med. *6*, 385–392.
Grootveld, M. and Halliwell, B. (1986) Biochem. J. *237*, 499.
Grootveld, M. and Halliwell, B. (1988) Biochem. Pharmacol. *37*, 271.
Halliwell, B. (1978) FEBS Lett. *98*, 321.
Halliwell, B., Grootveld, M. and Gutteridge, J.M.C. (1988) Method Biochem. Anal. *33*, 89–90.
Halliwell, B., Gutteridge, J.M.C. and Aruoma, O.I. (1987) Anal. Biochem. *165*, 215–219.
Halliwell, B. and Gutteridge, J.M.C. (1981) FEBS Lett. *128*, 347.
Halliwell, B., Kaur, H. and Ingelman-Sundberg. M (1991) Free Radical Biol. Med. 10, 439–441.
Halpen, H.J., Spencer, D.P., Van Polen, J., Bowman, M.K., Nelson, A.-C., Dowey, E.M. and Teicher, B.A. (1989) Rev. Sci. Instrum. *60*, 1040.
Hamers, M.N. and Roos, D. (1985) In: Oxidative Stress (Sies, H., ed.), Academic Press, New York, pp. 320–347.
Hildebrand, K. and Schulte-Frohlinde, D. (1989) Int. J. Radiat. Biol. *55*, 725.
Holt, A.J. (ed.) (1989) Advanced EPR. Elsevier, Amsterdam.
Holzman, J.L. (ed.) (1984) Spin Labelling in Pharmacology. Academic Press, New York.
Janco, R.L. and English, D. (1983) J. Lab. Clin. Methods *102*, 890–898.
Jonah, E.D. (1988) In: Chemical Reactivity in Liquids (Moreau M. and Turg, P., eds.), Plenum Press, New York.

Keran, L. and Bowman, M.K. (1990) Modern Pulsed and Continuous-Wave Electron Spin Resonance. Wiley-Interscience, New York.

Khan, A.V. and Kasha, M. (1963) J. Chem. Phys. *39*, 2105.

Kiefer, J. (1990) Biological Radiation Effects, Springer-Verlag, Berlin.

Klebanoff, S.J. and Rosen, H. (1978) J. Exp. Med. *147*, 490–506.

La Cagnin, L.P., Connor, H.D., Mason, R.P. and Thurman, R.G. (1988) Mol. Pharmacol. *33*, 351.

Lai, E.K., McCay, P.B., Nogichi, T. and Fong, K.L. (1979) Biochem. Pharmacol. *28*, 2231.

Leyroyer, V., Werner, L., Shaugnessy, S.D., Goddard, G.J. and Orr, F.W. (1987) Cancer Res. *47*, 4771.

Loschen, G. Flohe, L. and Chance, B. (1971) FEBS Lett. *18*, 261–264.

Maehly, A.C. and Chance, B. (1954) Methods Biochem. Anal. *1*, 357–424.

Maly, F.E. Nakamura, M., Gauchat, J.F., Urwyler, A., Walker, G., Dahinden, C.A., Cross, A.R., Jones, O.T.G. and Weck, A.L. (1989) J. Immunol. *142*, 1260–1267.

Maples, K.R., Jordon, S.J. and Mason, R.P. (1988) Mol. Pharmacol. *33*, 349.

Marnett, L.J., Wlodawer, P. and Samuelsson, B. (1974) Biochem. Biophys. Res. Commun. *60*, 1286.

Mason, R.P. and Mottley, C. (1987) In: Electron Spin Resonance (Symonds, M.C.R., ed.), Royal Society of Chemistry, London, 10B, 185.

Mehta, S., Bashford, L., Knox, P. and Pasternak, C. (1985) Biochem. J. *227*, 99.

Merzbach, D. and Obedeaunu, N. (1975) J. Med. Microbiol. *8*, 375–384.

Neta, P., Fessenden, R.W. and Schuler, R.H. (1971) J. Phys. Chem. *75*, 1654.

Niwa, Y., Sakane, T., Miyachi, Y. and Ozaki, M. (1984) J. Clin. Microbiol. *20*, 837–842.

Miyachi, Y., Yoshioka, A., Imamura, S. and Niwa, Y. (1987) J. Clin. Lab. Immunol. *22*, 81–84.

Nohl, H. and Hegner, D. (1978) Eur. J. Biochem. *82*, 563–567.

Peerless, A.G. and Stiehm, E.R. (1979) Clin. Immunol. Immunopathol. *38*, 1.

Perkins, M.J. (1980) Adv. Phys. Org. Chem. *17*, 1.

Peterhans, E., Jungi, T.H., Burge, T.H., Grob, M. and Jorg, A. (1988) In: Free Radicals, Methodology and Concepts (Rice-Evans, C. and Halliwell, B., eds.), Richelieu Press, London. pp. 307–334.

Peterhans, E., Mundy, J. and Parish, C.R. (1980) Eur. J. Immunol. *10*, 477.

Poole, C.P. (1967) Electron Spin Resonance: a Comprehensive Treatise on Experimental Techniques. John Wiley & Sons, New York.

Qin, L., Tripathi, G.N.R. and Schuler, R.H. (1985) Z. Naturforsch. *40a*, 1026.

Ramasarma, T. (1982) Biochim. Biophys. Acta *694*, 69–93.

Repine, J., Eaton, J., Anders, M., Hoidal, J. and Fox, R. (1979) J. Clin. Invest. *64*, 1642–1651.

Richmond, R., Halliwell, B., Chauhan, J. and Dardre, A. (1981) Anal. Biochem. *118*, 328.

Roder, J.C., Helfland, S.L., Werkmeister, J., McGarry, R., Beaumont, T.J. and Duwe, A. (1982) Nature *298*, 569.

Rook, G.A.W., Steele, J., Umar, S. and Dockerell, H.M. (1985) J. Immunol. Methods *82*, 161–167.

Rosen, H. and Klebanoff, S.J. (1979) J. Clin. Invest. *64*, 1725–1729.

Sangerhofer, A., Massotti, R.J. and Bowman, K. (1988) Israel J. Chem. *28*, 227.

Sagone, A.L., Decker, M.A., Wells, R.M. and Democko, C. (1980) Biochim. Biophys. Acta *628*, 90–97.

Segal, A. and Jones, O.T.G. (1979) Nature *276*, 515–517.

Segal, A. and Levi, A.J. (1975) Clin. Exp. Immunol. *19*, 309–318.

Seliger, H.H. (1964) J. Chem. Phys. *40*, 3133.

Slater, T.F., Sawyer, B. and Strauli, U. (1963) Biochim. Biophys. Acta *77*, 383–393.

Smith, J.P. and Trifumac, A.D. (1981) J. Phys. Chem. *85*, 1645.

Sugioka, K. and Nagano, M. (1976) Biochem. Biophys. Acta *423*, 203.

Swallow, A.J. (1973) Radiation Chemistry. Longman, London.

Symons, M.C.R. (1978) Chemical and Biological Aspects of Electron Spin Resonance Spectroscopy, Van Norstrand Reinhold, London.

Tauber, A.I. and Babior, B.M. (1977) J. Clin. Invest. *60*, 374–379.

Tomasi, A., Albano, E., Biasi, F., Slater, T.F., Vannini, V. and Dianzani, M.U. (1985) Chem. Biol. Interact. *55*, 303.

Von Sonntag, C. (1987) The Chemical Basis of Radiation Biology, Taylor & Francis, London.

Washino, K. and Schnabel, W. (1982) Intern. J. Radiat. Biol. *41*, 271.

Wefers, H. and Sies, H. (1988) In: Free Radicals, Methodology and Concepts (Rice-Evans, C. and Halliwell, B., eds.), Richelieu Press, London. pp. 335–347.

Weiss, S.J., Rustagi, P.K. and LoBuglio, A.F. (1978) J. Exp. Med. *147*, 316–322.

Wertz, J.E. and Bolton, J.R. (1972) Electron Spin Resonance, McGraw-Hill, New York.

Whitburn, K.D., Shieh, J.J., Sellers, R.M. and Hoffman, M.J. (1982) J. Biol. Chem. *257*, 1860.

Wrogemann, K., Weidemann, M.J., Peskar, B.A., Staudinger, H., Rietschel, E.T. and Fischer, H. (1978) Eur. J. Immunol. *8*, 749.

Yonetani, T. (1965) J. Biol. Chem. *240*, 4509–4514.

Yonetani, T. and Ray, G. (1965) J. Biol. Chem. *240*, 4503–4508.

Chapter 4

Bray, W.G. and Goria, A. (1932) J. Am. Chem. Soc. *54*, 2134.

Bray, R.C., Mautner, G.M., Fielden, E.M. and Carle C.I. (1977) In: Superoxide and Superoxide Dismutases (Michelson, A.M., McLoud, S.M. and Fridovich, I.A., eds.), Academic Press, pp. 395–412, London.

Carrington, A., Ingram, D.J.E., Lott, K.A.K., Schonland, D.S. and Symons, M.C.R.S. (1960) Proc. R. Soc. A. *254*, 1013.

Ceriotti, F. and Ceriotti, G. (1980) Clin. Chem. *26*, 327.

Davies, M.J.D. (1990) Free Rad. Res. Commun., in press.

Dixon, W.T. and Norman, R.O.C. (1962) *196*, 891.

Duffy, J.R. and Gandin, J. (1977) Clin. Biochem. *10*, 122–123.

Felton, R.H., Thomas, A.Y., Yu, N.-T. and Schonbaum, G.R. (1976) Biochim. Biophys. Acta *434*, 82–89.

Fenton, H.J.H. (1894) J. Chem. Soc. *65*, 899.

George, P. and Irvine, D.H. (1952) Biochem. J. *52*, 511–515.

Gilbert, G.C. and Jeff, M. (1988) In: Free Radicals: Chemistry, Pathology and Medicine (Rice-Evans, C. and Dormandy, T., eds.), Richelieu Press, London, pp. 25–49.

Gilbert, B.C. and Stell, P. (1990) J. Chem. Soc., Faraday Trans., in press.

Gutteridge, J.M.C. and Halliwell, B. (1987) Life Chem. Rep. *4*, 113.

Gutteridge, J.M.C., Rowley, D.A. and Halliwell, B. (1981) Biochem. J. *199*, 263.

Haber, F. and Weiss, J.J. (1934) Proc. R. Soc. Ser. A *147*, 332.

Halliwell, B., Aruoma, O.I. Mufti, G. and Bomford, A. (1988) FEBS Lett. *241*, 202.

Hartley, A. and Rice-Evans, C. (1989) Biochem. Soc. Trans. *17*, 488–489.

Hartley, A., Davies, M.J. and Rice-Evans, C. (1990) FEBS Lett. *264*, 145–148.

Johnson, G.R.A., Nazhat, N.B. and Saadalla-Nazhat, R.A. (1988) J. Chem. Soc. Faraday Trans. *1*, 501.

King, N.K., Looney, F.D. and Winfield, M.E. (1967) Biochim. Biophys. Acta *133*, 65–82.

Maeda, Y., Morita, Y. and Yoshida, C. (1973) J. Biochem. Tokyo *70*, 509–514.

Morehouse, K.M. and Mason, R.P. (1988) J. Biol. Chem. *263*, 1204.

Morrison, G. (1965) Anal. Biochem. *9*, 1124–1126.

Ortiz de Montellano P. and Kerr, D.E. (1983) J. Biol. Chem. *258*, 10558–10563.

Peters, S.W., Jones, B.M., Jacobs, A. and Wagstaff, M. (1985) In: Proteins of iron storage and transport (Spik, G., ed.), Elsevier, Amsterdam. p. 321.

Petersen, R., Symons, M.C.R. and Taiwo, F.A. (1989) J. Chem. Soc. Faraday Trans. *85*, 2435–2444.

Rakshit, G., Spiro, T.G. and Uyeda, M. (1976) Biochem. Biophys. Res. Commun. *71*, 803–808.

Rao, D.N.R., Flitter, W.D. and Mason, R.P. (1988) Cellular Antioxidant Defence Mechanisms (Chow, C.H., ed.), CRC Press, Boca Raton, FL, pp. 60–71.

Rice-Evans, C. and Baysal, E. (1987) Biochem. J. *244*, 191–196.

Turner, J.J.O., Rice-Evans, C., Davies, M.J. and Newman, E.S.R. (1991) Biochem. J. 1991, in press.

Whitburn, K.D. (1987) Arch. Biochem. Biophys. *253*, 419–430.

Whitburn, K.D. (1988) Arch. Biochem. Biophys. *267*, 614–622.

Winyard, P.G., Blake, D.R., Chirico, S. and Gutteridge, J.M.C. (1987) Lancet *i*, 69–72.

Witt, S.N. and Chan, S.I. (1987) J. Biol. Chem. *262*, 1446–1448.

Yamada, H. and Yamazaki, I. (1974) Arch. Biochem. Biophys. *165*, 728–738.

Yonetani, T. and Scleyer, H. (1967) J. Biol. Chem. *242*, 1974–1979.

Chapter 5

Benedetti, A., Casini, A.F., Ferrali, M. and Comporti, M. (1979) Biochem. Pharmacol. *28*, 2909–2918.

Benedetti, A., Comporti, M. and Esterbauer, H. (1980) Biochim. Biophys. Acta *620*, 281–296.

Benedetti, A. Comporti, M., Fulceri, R. and Esterbauer, H. (1984) Biochim. Biophys. Acta *792*, 172–181.

Benedetti, A., Esterbauer, H., Ferrali, M., Fulceri, R. and Comporti, M. (1982) Biochim. Biophys. Acta *711*, 345–356.

Benedetti, A., Fulceri, R. and Comporti, M. (1984) Biochim. Biophys. Acta *793*, 489–493.

Benedetti, A., Pompella, A., Fulceri, R., Romani, A. and Comporti, M. (1986) Biochim. Biophys. Acta *876*, 658–666.

Bird, R.P. and Draper, H.H. (1984) Methods Enzymol. *105*, 299–305.

Bird, R.P., Hung, S.S.O., Hadley, M. and Draper, H.H. (1983) Anal. Biochem. *128*, 240–244.

Bligh, E.C. and Dyer, W.I. (1959) Can. J. Biochem. Physiol. *37*, 911–917.

Bond, A.M., Deprez, P.P., Jones, R.D., Wallace, G.G. and Briggs, M.H. (1980) Anal. Chem. *52*, 2211–2212.

Buege, J.A. and Aust, S.D. (1978) Methods Enzymol. *52C*, 302–310.

Buffinton, G.D., Cowden, W.B., Hunt, D.H. and Clark, I.A. (1986) Aust. NZ. J. Med. *16*, 179.

Burk, R.F. and Ludden, T.M. (1989) Biochem. Pharmacol. *38*, 1029–1032.

Buttriss, J.L. and Diplock, A.T. (1988) Biochim. Biophys. Acta *963*, 61–69.

Cathcart, R., Schwiers, E. and Ames, B.N. (1983) Anal. Biochem. *134*, 111–116.

Cathcart, R., Schwiers, E. and Ames, B.N. (1984) Methods Enzymol. *105*, 352–358.

Cawood, P., Wickens, D.G., Iversen, S.A., Braganza, J.M. and Dormandy, T.L. (1983) FEBS Lett. *162*, 239–243.

Chan, H.W.-S. and Coxon, D.T. (1987) In: Autoxidation of unsaturated lipids. (Chan, H.W.-S., ed.), Academic Press, New York, pp. 17–50.

Cheeseman, K.H., Emery, S., Maddix, S.P., Slater, T.F., Burton, G.W. and Ingold, K.U. (1988) Biochem. J. *250*, 247–252.

Chiang, S.P., Gessert, C.F., and Lowry, O.H. (1957) In: Research Report 56–113, Air University School of Aviation Medicine USAF, Texas, p. 1.

Christie, W.W. (1982a) Lipid Analysis. Pergamon Press Ltd. Oxford, England

Christie, W.W. (1982b) J. Lipid Res. *23*, 1072–1075.

Christie, W.W. (1987) High-performance liquid chromatography and lipids. Pergamon Press Ltd., Oxford, England.

Cohen, G. (1982) In: Lipid Peroxides in Biology and Medicine (Yagi, K., ed.), Academic Press, New York, pp. 199–211.

Crump, B.J., Thurnham, D.I., Situnayake, R.D. and Davis, M. (1985) Lancet *ii*, 955–956.

Csallany, A.S., Guan, M.D. Manwaring, J.D. and Addis, P.B. (1984) Anal. Biochem. *142*, 277–283.

Curzio, M., Poli, G., Esterbauer, H., Biasi, F., Mauro, C.D.I. and Dianzani, M. (1986) IRCS Med. Sci. *14*, 984–985.

Dillard, C.J., Dumelin, E.E. and Tappel, A.L. (1977) Lipids *12*, 109–122.

Edisbury, J.R., Morton, R.A. and Lovern, J.A. (1933) Biochem. J. *27*, 1451–1460.

Erickson, B.W. (1974) Org. Synth. *54*, 19–27.

Erskine, Y.J., Iversen, S.A. and Davies, R. (1985) Lancet *i*, 554–555.

Esterbauer, H. (1982) In: Free Radicals, Lipid Peroxidation and Cancer (McBrien, D.C.M., and Slater, T.F., eds.), Academic Press New York, pp. 101–128.

Esterbauer, H. and Slater, T.F. (1981) IRCS. Med. Sci. *9*, 749–750.

Esterbauer, H., Cheeseman, K.H., Dianzani, M.U., Poli, G. and Slater, T.F. (1982) Biochem. J. *208*, 129–140.

Esterbauer, H., Jurgens, G., Quchenberger, O. and Koller, E. (1987) J. Lipid. Res. *28*, 495–509.

Esterbauer, H., Lang, J., Zadravec, S. and Slater, T.F. (1984) Methods Enzymol. *105*, 319–328.

Esterbauer, H. and Weger, W. (1967) Mh. Chem. *98*, 1994–2000.

Esterbauer, H., Zollner, H. and Schaur, R.J. (1988) ISI Altas. Sci. Biochem. 1 311–317.

Evans, C.D., List, G.R., Dolev, A., McConnell, D.G. and Hoffmann, R.F. (1969) Lipids *2*, 432.

Fairbank, J., Ridgway, L., Griffin, J., Wickens, D., Singer, A. and Dormandy, T.L. (1988) Lancet *ii*, 329.

Farmer, E.H. and Sundralingham, A. (1942) J. Chem. Soc. pp. 121–129.

Farmer, E.H., Bloomfield, G.F., Sundralingham, A. and Sutton, D.A. (1942) Trans. Faraday. Soc. *38*, 348–356.

Fink, R., Marjot, D.H., Clemens, M.R., Patsalos, P., Cawood, P., Iversen, S.A. and Dormandy, T.L. (1985) Lancet *ii*, 291–294.

Fraga, C.G., Leibovitz, B.E. and Tappel, A.L. (1988) Free Rad. Biol. Med. *4*, 155–161.

Frankel, E.N. and Neff, W.E. (1983) Biochim. Biophys. Acta *754*, 264–270.

Frei, B., Yamamoto, Y., Niclas, D., Stocker, R., Cross, C.E. and Ames, B.N. (1988) In: Free Radicals, Methodology and Concepts. (Rice-Evans, C. and Halliwell, B., eds.) Richelieu, London, pp. 349–368.

Frommer, U., Ulrich, V. and Staudinger, H. (1970) Hoppe-Seyler's. Z. Physiol. Chem. *351*, 903–912.

Gebicke, J.M. and Guille, J. (1989) Anal. Biochem. *176*, 360–364.

Gilbert, H.S., Stump, D.D. and Roth, E.F. (1984) Anal. Biochem. *137*, 282–286.

Gillam, A.E., Heilbron, I.M., Hilditch, T.P. and Morton, R.A. (1931) Biochem. J. *25*, 30–38.

Graff, A. (1982) Methods Enzymol. *86*, 386–391.

Gree, R., Tourbah, H. and Carrie, R. (1986) Tetrahedr. Lett. *27*, 4983–4986.

Griffin, J.F.A., Wickens, D.G., Tay, S.K., Singer and Dormandy, T.L. (1987) Clin. Chim. Acta *163*, 143–148.

Gutteridge, J.M.C. and Tickner, T.R. (1978) Anal. Biochem. *91*, 250–257.

Harrison, K., Cawood, P., Iversen, A. and Dormandy, T.L. (1985) Life Chem. Rep. *3*, 41–44.

Hicks, M. and Gebicki, J.M. (1979) Anal. Biochem. *99*, 249–254.

Hornstein, M. and Crowe, H.W. (1962) Anal. Chem. *34*, 1037–1038.

Iversen, S.A., Cawood, P. and Dormandy, T.L. (1985) Ann. Clin. Biochem. *22*, 137–140.

Iversen, S.A., Cawood, P., Madigan, M.J., Lawson, A.M. and Dormandy, T.L. (1984) FEBS Lett. *171*, 320–324.

Iversen, S.A., Cawood, P., Madigan, M.J., Lawson, A.M. and Dormandy, T.L. (1985) Life. Chem. Rep. *3*, 45–48.

Jordan, R.A. and Schenkman, J.B. (1982) Biochem. Pharmacol. *31*, 1393–1400.

Kappus, H. and Muliawan, M. (1988) Biochem. Pharmacol. *31*, 597–600.

Keston, A.S. and Brandt, R.B. (1965) Anal. Biochem. *11*, 1–5.

Kohn, H.I. and Liversedge, M. (1944) J. Pharmacol. Exp. Ther. 82, 292–294.
Kostrucha, J. and Kappus, H. (1986) Biochim. Biophys. Acta 879, 120–125.
Kosugi, H. and Kikugawa, K. (1986) Lipids 21, 537–542.
Kukuda, Y., Stanley, D.W. and Van de Voort, F.R. (1981) J. Am. Oil. Chem. Soc. 58, 773.
Kwon, T.W. and Watts, B.M. (1963) Anal. Chem. 35, 733–737.
Kwon, T.W. and Watts, B.M. (1963) J. Food. Sci. 28, 627–630.
Lands, W.E.M. (1988) In: Free Radicals Methodology and Concepts (Rice-Evans, C. and Halliwell, B., eds.) Richelieu Press, London p. 218.
Lawrence, G.D. and Cohen, G. (1984) Methods Enzymol. 105, 305–311.
Lea, C.H. (1931) Proc. R. Soc. Lond. B. 108, 175–189.
Lea, C.H. (1952) J. Sci. Food. Agric. 3, 586–594.
Lee, H.S. and Csallany, A.S. (1987) Lipids 22, 104-107.
Lemoyne, M., Van Gossum, A., Kurian, R., Ostro, M., Axler, J. and Jeejeebhoy, K.N. (1987) Am. J. Clin. Nutr. 46, 267–272.
Lynch, C., Lim, C.K., Thomas, M. and Peters, J.J. (1983) Clin. Chim. Acta 13, 117–122.
Lynch, R.D., Schneeberger, E.E and Geyer, R.P. (1976) Biochemistry 15, 193–200.
Manwaring, J.D. and Csallany, A.S. (1981) J. Nutr. 111, 2172–2179.
Manwaring, J.D. and Csallany, A.S. (1988) Lipids. 23, 651–655.
Marshall, P.J., Warso, M.A. and Lands, W.G.M. (1985) Anal. Biochem. 145, 192–199.
Mopper, K., Stahovec, W.L. and Johnson, L. (1983) J. Chrom. 256, 243–252.
Morrison, M.R. and Smith, L.M. (1964) Lipid. Res. 5, 231–239.
Muller, A. and Sies, H. (1984) Methods Enzymol. 105, 311–319.
Nair, V. and Turner, G.A. (1984) Lipids. 19, 804–805.
Pendleton, R.B. and Lands, W.E.M. (1987) Free Rad. Biol. Med. 3, 337–339.
Petonen, K., Ffaffli, P. and Itkonen, A. (1984) J. Chromatog. 315, 412–416.
Poli, G., Dianzani, M.U., Cheeseman, K.H., Slate, T.F., Lang, J. and Esterbauer, H. (1985) Biochem. J. 227, 629–638.
Pompella, A., Romani, A., Fulceri, R., Benedetti, A. and Comporti, M. (1988) Biochim. Biophys. Acta 96, 293–298.
Porter, N.A. (1984) Methods Enzymol. 105, 273–283.
Pryor, W.A. and Castle, L. (1984) Methods Enzymol. 105, 293–299.
Pryor, W.A., Stanley, J.P. and Blair, E. (1976) Lipids. 11, 370–379.
Ramenghi, U., Chiarpotto, E., David, O., Miniero, M., Piga, A., Cecchini, G.., Biasi, F., Poli, G. and Esterbauer, H. (1985) IRCS. Med. Sci. 13, 273–274.
Recknagel, R.O. and Glende, E.A. (1984) Methods Enzymol. 105, 331–337.
Riely, C.A., Cohen, G. and Lieberman, M. (1974) Science 183, 208–210.
Reiter, R. and Burk, R.F. (1987) Biochem. Pharmacol. 36, 925–929.
Rice-Evans, C., McCarthy, P.T., Hallinan, T., Green, N.A., Gor, J. and Diplock, A.T. (1989) Free Rad. Res. Commun. 7, 307–313.
Singer, A. Tay, S.K., Griffin, J.F.A., Wickens, D.G. and Dormandy, T.L. (1987) Lancet i, 537–538.
Sinnhurber, R.O. and Yu, T.C.E. (1958) Food Technol. 12, 9–12.
Sinnhurber, R.O. and Yu, T.C.E. (1958) Food Res. 23, 626–633.
Slater, T.F. (1988) In: Free Radicals, Methodology and Concepts (Rice-Evans, C. and Halliwell, B., eds.), Richelieu, London, pp. 347—368.

275

Slater, T.F. and Sawyer, B.C. (1971) Biochem. J. *123*, 805 814.

Smith, C.V. and Anderson, R.E. (1987) Free Rad. Biol. Med. *3*, 341–344.

Suzuki, Y. (1985) Bonseki, Kagaku, *34*, 314.

Szebeni, J., Eskelson, C., Sampliner, R., Hartmann, B., Griffin, J., Dormandy, T.L. and Watson, R.R. (1986) Alcoholism Clin. Exp. Res. *10*, 647–850.

Tappel, A.L. (1980) In: Free Radicals in Biology IV. (Pryor, W.A., ed.), Academic Press, New York, pp. 2–47.

Tay, S.K., Singer, A., Griffin, J.F.A., Wickens, D.G. and Dormandy, T.L. (1987) Clin. Chim. Acta. *163*, 149–152.

Tokarz, A., McCarthy, P.T. and Diplock, A.T. (1988) In: Free Radicals: Chemistry, Pathology and Medicine (Rice-Evans, C. and Dormandy, T., eds.), Richelieu, London, p. 151.

Tomita, I., Yoshino, K. and Sano, M. (1987) In: Clinical and Nutritional Aspects of Vitamin E. (Hayaishi, O. and Mino, M., eds.), Elsevier, Amsterdam, pp. 277–280.

Uchiyama, M. and Mihara, M. (1988) Anal. Biochem. *86*, 271–278.

Van Gossum, A., Shariff, R., Lemoyne, M., Kurian, R. and Jeejeebhoy, K. (1988) Am. J. Clin. Nutr. *48*, 1394–1399.

Wendel, A. (1987) Free Rad. Biol. Med. *3*, 355–358.

Wickens, D.G. and Dormandy, T.L. (1988) In: Free Radicals: Chemistry, Pathology and Medicine (Rice-Evans, C. and Dormandy, T., eds.) Richelieu Press, London, pp. 237–252.

Willson, R.L. (1979) In: Oxygen Free Radicals and Tissue Damage. Ciba Foundation Symposium 65. Excerpta Medica, Amsterdam, p. 19.

Winkler, P., Lindler, W., Esterbauer, H., Schauenstein, E., Schaur, R.J. and Khoschsorur, G.A. (1984) Biochim. Biophys. Acta *796*, 232–237.

Yagi, K. (1984) Methods Enzymol. *105*, 328–331.

Yamamoto, Y. and Ames, B.N. (1987) Free Rad. Biol. Med. *3*, 359–361.

Yoshino, K., Sano, M., Fujita, M. and Tomita, I. (1986), Chem. Pharm. Bull. *34*, 5184–5187.

Yoshino, K., Sano, M., Fujita, M. and Tomita, I. (1986a) Chem. Pharm. Bull. *34*, 5184–5187.

Yoshino, K., Matsuvra, T., Sano, M., Saito, S.I. and Tomita, I. (1986b) Chem. Pharm. Bull. *34*, 1694 1700.

Yost, R., Stoveken, J. and Maclean, W. (1977) J Chromatogr. *134*, 73–75.

Yu, L.W., Latriano, L., Duncan, S., Hartwick, R.A. and Witz, A. (1986) Anal. Biochem. *156*, 326–337.

Zarling, E.J. and Clapper, M. (1987) Clin. Chem. *33*, 140–141.

Chapter 6

Aebi, H. (1984) Methods Enzymol. *103*, 121.

Alin, P., Jennson, H., Gutenberg, C., Danielson, U.H., Tahir, M.K. and Manervic, B. (1985) Anal. Biochem. *146*, 313.

Anderson, S.M. and Krinsky, N.I. (1973) Photochem. Photobio. *18*, 403.

Bieri, J.G., Tolliver, T.J. and Catagnani, G.L. (1979) Am. J. Clin. Nutr. *32*, 2143.
Bunnell, R.H. (1971) *6*, 245.
Buttriss, J.L. and Diplock, A.T. (1984) Methods Enzymol. *103*, 131.
Coker, O.E. (1978) An investigation into selenium metabolism with reference to the interactions between vitamin E, selenium and other trace elements. Ph. D. Thesis, University of London.
Connett, J.E., Kuller, L.H., Kielsberg, M.O., Polk, B.F., Collins, G., Rider, A. and Hulleys, B. (1989) Cancer *64*, 126.
Davies, B.H. (1976) In: Chemistry and Biochemistry of Plant Pigments. (Goowin, T.W., ed.), Academic Press, New York, p. 38.
De Leenheer, A.P., De Bevere, V.O., De Ruyter, M.G.M. and Claeys, A.E. (1979) J. Chromatog. *162*, 408.
De Ritter, E. and Purcell, A.E. (1981) In: Carotenoids as Colorants and Vitamin A Precursors (Bassenfeind, J.C., ed.), Academic Press, New York, p. 815.
Flohe, L. and Gunzler, W.A. (1984) Methods Enzymol. *105*, 114.
Flohe, L. and Otting, F. (1984) Methods Enzymol. *103*, 93.
Gutenberg, C. and Mannervik, B. (1979) Biochem. Biophys. Res. Commun. *86*, 1304.
Habig, W.H. and Jakoby, W.B. (1981) Methods Enzymol. *77*, 398.
Krinsky, N.I. and Deneke, S.M. (1982) J. Natl. Cancer. Inst. *69*, 205.
Krinsky, N.I. and Welankiwar, S. (1984) Methods Enzymol. *105*, 155.
Laemmli, U.K. (1970) Nature, *227*, 680–685.
McCord, J.M. and Fridovich, I. (1969) J. Biol. Chem. *244*, 6049.
McMurray, C.H. and Blanchflower, W.J. (1979) J. Chromatog. *178*, 525.
Macpherson, A., Sampson, A.K. and Diplock, A.T. (1988) Analyst *113*, 281–283.
Metcalfe, T., Bowen, D.M. and Muller, D.P.R. (1989) Neurochem. Res. *14*, 1209.
Olson, O., Palmer, I.S. and Cary, E.E. (1975) J. Assoc. Off. Anal. Chem. *58*, 117.
Paglia, D.E. and Valentine, W.N. (1967) J. Lab. Clin. Med. *70*, 159.
Vatassery, G.T. and Hagen, W.J. (1979) J. Chromatog. *178*, 525.

Chapter 7

Alvarez, J., Haris, P.I., Lee, D.C. and Chapman, D. (1987) Biochim. Biophys. Acta *916*, 5.
Alvarez, J., Lee, D.C., Baldwin, S.A. and Chapman, D. (1987) J. Biol. Chem. *262*, 3502.
Bump, E.A., Yu, N.Y. and Brown, J.M. (1982) Science *127*, 544–565.
Chance, B. (1981) Curr. Top. Cell Reg. *18*, 343–360.
Chapman, D., Gomez-Fernandez, J.C., Gomi, F.M. and Barnard, M. (1960) J. Biochem. Biophys. Methods *2*, 315.
Deuticke, B., Henseleit, U., Haest, C.W.M., Heller, K.B. and Dubbelman, T.M.A.R. (1989) Biochim. Biophys. Acta *982*, 53–61.
Ellman, G. (1959) Arch. Biochem. Biophys. *82*, 70–77.
Grassetti, D.R. and Murray, J.F. (1967) Arch. Biochem. Biophys. *199*, 41–49.
Griffiths, P.R. (1980) In: Analytical Applications of FTIR to Molecular and Biological Systems (Durig, J.R., ed.), Reidel, Holland. p. 11.

Gutteridge, J.M.C. and Wilkens, S. (1983) Biochim. Biophys. Acta *759*, 38–41.

Habeeb, A.F.S.A. (1966) Biochim. Biophys. Acta *121*, 440–454.

Haest, C., Plasa, G., Kamp, D. and Deuticke, B. (1978) Biochim. Biophys. Acta *509*, 21–32.

Haris, P.I., Rice-Evans, C., Ahmad, J., Khan, R., Chapman, D. (1989) In: Free Radicals, Diseased States and Anti-Radical Interventions (Rice-Evans, C., ed.), Richelieu Press, London. pp. 307–353.

Haris, P.I., Lee, D.C. and Chapman, D. (1986) Biochim. Biophys. Acta *874*, 255.

Jones, L.A., Holmes, J.C. and Seligman, R.B. (1956) Anal. Chem. *28*, 191–198.

Karam, L.R. and Simic, M.G. (1988) Anal. Chem. *60*, 1117.

Koenig, J.L. and Tabb, P.L. (1980) In: Analytical Applications of FTIR to Molecular and Biological Systems (Durig, J.R., ed.), Reidel, Holland, pp. 241–267.

Lappin, R. and Clark, C.L. (1951) Anal. Biochem. *23*, 541–542.

Lee, D.C. and Chapman, D. (1986) Biosci. Rep. *6*, 235.

Meister, A., Anderson, M.E. (1983) Annu. Rev. Biochem. *52*, 711–760.

Mitchell, J.B. and Russo, A. (1983) Radiat. Res. *95*, 471–485.

Oliver, C.N., Ahn, B., Moerman, E., Goldstein, S. and Statman, E.R. (1987) J. Biol. Chem. *262*, 5488–5491.

Pirie, A. (1972) Biochem. J. *128*, 1365–1367.

Prutz, W.A., Butler, J. and Land, E.J. (1983) Int. J. Radiat. Biol. *44*, 183–196.

Russo, A. and Bump, E.A. (1989) Methods Biochem. Anal. *33*, 165–241.

Sies, H. and Akerboom, P.M. (1984) Methods Enzymol. *105*, 445–451.

Susi, H. and Byler, D.M. (1986) Methods Enzymol. *130*, 290.

Teale, F.W.J. (1960) Biochem. J. *76*, 381–388.

Udenfried, S., Stein, S., Bohlen, P., Daiman, W., Leimgruber, W. and Wiegele, M. (1972) Science *178*, 871.

Wills, K.J. and Teale, F.W.J. (1986) Anal. Biochem. *158*, 336–448.

Chapter 8

Adams, R.L.P. (1980) Cell Culture for Biochemists, Vol. 8. In: Laboratory Techniques in Biochemistry and Molecular Biology (Work, T.S. and Burdon, R.H., eds.), Elsevier, Amsterdam.

Aruoma, O.I., Halliwell, B. and Dizdaroglu, M. (1989) J. Biol. Chem. *264*, 13024–13028.

Aruoma, O.I., Halliwell, B., Gajewski, E. and Dizdaroglu, M. (1989) J. Biol. Chem. *264*, 20509-20512.

Boffey, S.A. (1983) In: Techniques in Molecular Biology (Walker J.M. and Gastra, W., eds.), Croom Helm, London, pp. 257–272.

Boon, P.J., Cullis, P.M., Symons, M.C.R. and Wren, B.W. (1984) J. Chem. Soc., Perkin. Trans 2, 1393.

Bradley, M.O. and Kohn, K.W. (1979) Nucleic Acids Res. *7*, 793–804.

Britten, R.J. and Kohne, D.E. (1968) Science *161*, 529–535.

Burkhoff, A.M. and Tullius, T.D. (1987) Cell *48*, 935–943.

Chalekapis, G. and Beato, M. (1989) Nucleic Acids Res. *17*, 1783.

Collins, A.R.S. (1977) Biochim. Biophys. Acta *478*, 461–473.

Cook, P.R. and Brazell, I.A. (1976) Nature *263*, 677–681.

Cullis, P.M., Jones, G.D.D. and Symons, M.C.R. (1987) Nature *330*, 773.

Davis, L.G., Dibner, M.D. and Battey, J.E. (1986) Basic Methods in Molecular Biology. Elsevier, New York.

Dizdaroglu, M. (1985) Anal. Biochem. *144*, 593–603.

Dizdaroglu, M. (1986) Biotechniques *4*, 536–546.

Dizdaroglu, M. and Bergtold, D.S. (1986) Anal. Biochem. *156*, 182.

Dizdaroglu, M. (1988) In: Free Radicals: Methodology and Concepts (Rice-Evans, C. and Halliwell, B., eds.), Richelieu Press, London. pp. 123–138.

Gardiner, K., Laas, W. and Patterson, D. (1986) Somat. Cell Mol. Genet. *12*, 185.

George, A.M., Sabovlev, S.A., Hart, L.E., Cramp, W.A., Harris, G. and Hornsey, S. (1987) Br. J. Cancer *55*, 141.

Gidoni, D., Kadonaga, J.T., Barrera-Saldana, H., Takahaski, K., Chambon, P. and Tjian, R. (1985) Science *230*, 516–519.

Hayes, J.J. and Tullius, T.D. (1989) Biochemistry *28*, 9521–9527.

Kadonaga, J.T., Jones, K.A. and Tjian, R. (1986) Trends Biochem. Sci. *11*, 21–23.

Kohn, K.W. (1979) In: Methods in Cancer Research, Vol. XVI (Devita, V.T. and Bosch, J., eds.), Academic Press, London. p. 291.

Lu, M., Guo, Q., Wink, D.K. and Kallenbach, N.R. (1990) Nucleic Acids Res. *18*, 3333–3337.

Meneghini, R. (1976) Biochim. Biophys. Acta *425*, 419–427.

Mullinger, A.M. and Johnson, R.T. (1985) J. Cell. Sci. *73*, 159–186.

Schickov, P., Metzger, W., Werel, W., Lederer, H. and Heumann, H. (1990) EMBO J. *9*, 2215–2220.

Studier, F.W. (1965) J. Mol. Biol. *11*, 373–390.

Tullius, T.D. and Dombroski, B.A. (1986) Proc. Natl. Acad. Sci. USA *83*, 5469–5473.

Ueda, K., Kobayashi, S., Morita, J. and Komano, T. (1985) Biochim. Biophys. Acta *824*, 341–348.

Von Sonntag, C. (1987) The Chemical Basis of Radiation Biology, Taylor & Francis, London.

Zorbas, H., Rogge, L., Meisterernst, M. and Winnacker, E.L. (1989) Nucleic Acids Res. *17*, 7735–7748.

Subject index

284